Agricultural Biotechnology Sector
Issues Impacting Innovations

Agricultural Biotechnology Sector
Issues Impacting Innovations

Mandeep Kaur

RANDOM PUBLICATIONS
NEW DELHI (INDIA)

Agricultural Biotechnology Sector: Issues Impacting Innovations

ISBN 978-93-5111-111-5

Published in 2013 in India by

RANDOM PUBLICATIONS

4376-A/4B, Gali Murari Lal, Ansari Road
New Delhi-110 002
Phone : +9111-43580356, 011-23289044
e-mail: randompublications@hotmail.com

Type Setting by : Friends Media, Delhi-110089
Printed at : Thomson Press (India) Ltd.

Contents

Preface

Agricultural biotechnology is a collection of scientific techniques, including genetic engineering, that are used to create, improve, or modify plants, animals and microorganisms. Using conventional techniques, such as selective breeding, scientists have been working to improve plants and animals for human benefit for hundreds of years. In fact, all agricultural biotechnologies should be looked upon as an extension and integral part of traditional breeding and agriculture, contributing successfully to shortening the breeding cycle, and to production of quality agricultural commodities.

The strategic integration of biotechnology tools into India agricultural systems can revolutionise Indian farming and usher in a new era. Genetic engineering is clearly the most revolutionary tool to impact agricultural research since the discovery of genetics by Mendel. Prior to genetic engineering, the exchange of DNA material was possible only between individual organisms of the same species.

Every year, millions of people worldwide die from starvation and nutritional deficiencies. Although these challenges remain significant, we are making some progress. Agricultural technologies have increased caloric intake on a per-capita basis in nearly every nation of the world. technological development— particularly in the area of agricultural biotechnology—is crucial if we want to further reduce malnutrition and starvation and meet the needs of growing populations. Unfortunately, misperceptions about biotechnology have led to activist movements to slow down, if not eliminate, this critical technology via government regulation. In the United States, the advancement of biotech crops has slowed down because of attempts to regulate them differently than conventionally grown foods.

This book sums up information about agricultural bio-technology and is a valuable tool for students as well as teachers.

—*Author*

1

Introduction

In the last ten years the seed and pesticide industries have undergone a substantial number of structural changes. These changes are due to a number of factors, some of which are common to all industries and some of which are specifically tied to the biotechnology that is increasingly important in the seed and chemical industries.

The focus of this paper is on these latter linkages. The horizontal mergers and acquisitions can be linked to R&D costs, economies of scale and scope created by intellectual property rights, and to regulatory costs, while the increased vertical linkages are connected to product complementarity and to the difficulty in enforcing certain types of intellectual property.

For example, Cornell University researchers are developing a rice variety that yields 20 percent to 40 percent more food per acre and a tomato that produces 48 percent higher yields per acre. Biotechnology promises to play a critical role in reducing blindness caused by vitamin A deficiencies.

Every year in Southeast Asia, one quarter of a million children go blind because of a lack of vitamin A in their diets. Worldwide, 124 million children suffer from vitamin A deficiency. Increasing the intake of vitamin A could prevent one to two million deaths per year and eliminate numerous health problems.

Poverty and lack of infrastructure make the traditional methods of vitamin consumption (a varied diet or supplements) impossible. Agricultural biotechnology, however, can help to address the problem of childhood blindness and death by ensuring that people get this nutrient in the key staple of their diet, rice.

Researchers have spliced into the genome of rice, the beta-carotene gene from a daffodil to create "golden rice" — rice high in beta-carotene, which the human body converts into vitamin A. This is among the most well-known examples of biotechnology's benefits. Countless other genetically engineered crop varieties are being developed to address similar problems of poor nutrition or productivity.

ADVANCES BIOTECHNOLOGY

Since the discovery of recombinant DNA (rDNA) and the creation of the first transgenic organism over twenty years ago, biotechnology has shifted from the laboratory to the marketplace. Particularly in the West, biotechnology is a rapidly growing industry. Already there are almost 100 biotech medicines on the market, with another 350 in late-stage clinical trials. (Feldbaum, 2000) In purely market terms, 50 biotech companies had market capitalizations of more than US $1 billion by the end of 2000. (Van Brunt, 2001) And the completion of a preliminary map of the human genome has only served to fuel future hopes. In agriculture, the rapid of growth of the biotechnology industry has been equally impressive. By 2000, 16 per cent of area under cultivation worldwide was planted to GM crops. In just five years, the global area under transgenic cultivation increased more than 25 fold, from 2.8 million hectares in 1996 to 44.2 million hectares in 2000. More than three-quarters of all GM crops, however, are produced in the developed world. Four countries in particular account for 99 percent of the total area planted to GM crops: the US (with 68% of the world's total), Argentina (10%), Canada (7%), and China (1%), with nine countries comprising the remaining one percent.

HOW AGRICULTURAL BIOTECHNOLOGY IMPROVES PLANT SAFETY

Many biotechnology skeptics contend that such engineering is unsafe because it is "unnatural." Ironically, biotechnology is actually an improved method for plant breeding. Most of what we eat today has already undergone gene manipulation under the traditional breeding methods.

The traditional methods and biotechnology differ in that traditional methods are cruder because they do not give researchers control over all the genes they crossbreed. Some desirable characteristics might result with such methods, but undesirable ones might as well. It could take a long time of continual breeding to eliminate the most problematic traits and focus on the best ones. Compared to these more "traditional" methods of hybridization, genetic engineering is a wonder of precision and accuracy. Genetic engineering enables scientists to pinpoint the genes that express a desirable characteristic in one plant and to insert those genes into the genome of another plant. Scientists mark the insertion point so that they know where these novel genes are located in the genome of the new plant. While there are several different methods of genetically engineering plants,

EXACTLY BIOTECHNOLOGY

For centuries breeders have improved plants through selective breeding and crossbreeding different types of plants. Agricultural biotechnology — also referred to as "genetic engineering" or "genetic modification" (GM) — is a more sophisticated and precise means of modifying plant genes. Instead of transferring thousands of genes as traditional crossbreeding does, biotechnology enables the transfer of only a few select plant genes. Breeders essentially cut the desirable traits out of one plant's DNA and paste them into another — a process called "gene splicing." In addition, undesirable traits can be targeted and removed. Using biotechnology, plant breeders can produce better crops.

BIOTECHNOLOGY'S CRITICAL BENEFITS

More affordable, higher quality food surely benefits people in developed nations, but those benefits pale in comparison to what biotech promises for the developing world. Biotechnology promises to reduce world hunger and disease by improving local productivity by adapting crops to local climates and soils; increasing yield by making plants stronger and more pest-resistant; making plants more nutritious by creating plants with higher vitamin and protein content; and making produce more affordable on the world market.

GENERAL BIOTECHNOLOGY INDICATORS

Since 1999, organizations such as the Organization for Economic Co-operation and Development (OECD) and the Organization of American States (OAS) began to compile statistics to describe the situation of the countries that are developing biotechnologies. A general view shows that countries in Latin America and the Caribbean (LAC) have significant gaps with countries of the North, especially in patterns of patenting and licensing technologies, and that the transferal of biotechnologies is of vital importance, especially those biotechnologies developed for the agricultural sector.

In LAC countries, measurements mainly include the analysis of the capacity of groups and research centers, their working fields, the number of enterprises and the market perspectives, whereas the statistics of OECD measure mainly the private investment, the generation of investor's capital in terms of new companies, money and employment, and the obtainment of patents among other indicators, which are not so representative for us due to the differences in economies and development systems.

An indicator to analyze the state of biotechnology in the countries is a function of the investment made for its developments and products. The United States of America has a much higher investment tendency in relation to other countries. In 1997, the industry invested around US$ 7.300 millions, being 3.7% of the total investment for research and development (NSF, 2004), and as for 2001 it reported that investment of industry in research and development of biotechnology reached US$ 16.400 millions, representing 10% of the total investments in science and technology. European countries such as Germany, with more than US$ 1.000 millions (6.7% of total), United Kingdom with US$ 705 millions (7.8% of total) and France with US$ 560 millions having a 4.4% of the total of governmental investment in R&D, follow the list with the largest assigned budget for biotechnology according to statistics of OECD.

The situation in countries members of the OECD reveals a striking difference between USA and EU. According to data provided in 2001 by the consultant company Ernst and Young, USA investment in R&D was € 11.400 million (2.7% of DGP) whilst EU investment was € 4.977 million (1,8% of DGP). There are 1,262 companies in USA totalling 162,000 employees and 6.7% of them are involved in research, in contrast to 1,570 companies in EU with 61,000 employees, 2.5% of them conducting research.

In the compendium of biotechnology statistics of OECD, differences can also be observed. The first one is the amount of investment in risk capitals in biotechnology, which goes up to US$ 1.182 millions in the USA, and US$ 687 millions for the EU during 1999. The bioengineering products reached that year up to 1% of the total exports of USA with a total amount of US$ 1.340 millions. Its import of products from OECD countries, especially from Belgium, France, Switzerland and The Netherlands were that same year of US$ 970 millions. Its most important commercial partners where most of the exports are sent to include: Belgium, Japan, Canada and Germany.

This indicator allows analyzing the degree of technological delay resulting from the efforts of countries such as Colombia, where it represents approximately US$ 1.8 millions (constant Colombian pesos as in 2003. Exchange rate 1 US$ = $ 2.200 Col) and 3.7% of the total governmental investment of the National Biotechnology Program (NBP) made through Colciencias during 1997, and during the following years the tendency remained negative, reaching in 1999 almost US$ 780.000 and 4.8% (OCyT, 2004).

This situation somehow explains why Colombia reports 8 patents granted to research groups associated with the NBP (OCyT, 2004), whereas the USA already had more than 3.600 patents and Canada more than 500 during the year 2000 (Van Beuzekom, 2001). It could be said that the relation between the investment in biotechnology and the number of patents is more than US$ 2 million per patent in the USA, US$ 640.000 in Canada and US$ 22.000 in Colombia.

On the other hand, the development of biotechnology in LAC has been characterized by being a repetitive model where the research projects do not correspond to the real needs of production and food security, and has a low level of innovation and scientific creativity. Only 2% out of 30 scientific articles of high impact on biotechnology came from laboratories based within the region, despite LAC possessing important biodiversity resources to develop innovative products.

SIGNIFICANT OF AGRICULTURAL BIOTECHNOLOGY

Every year, millions of people worldwide die from starvation and nutritional deficiencies. Although these challenges remain significant, we are making some progress. Agricultural technologies have increased caloric intake on a per-capita basis in nearly every nation of the world. technological development— particularly in the area of agricultural biotechnology — is crucial if we want to further reduce malnutrition and starvation and meet the needs of growing populations. Unfortunately, misperceptions about biotechnology have led to activist movements to slow down, if not eliminate, this critical technology via government regulation. In the United States, the advancement of biotech crops has slowed down because of attempts to regulate them differently than conventionally grown foods. In Europe, antibiotech movements are much stronger, and international bureaucrats are pushing various international agreements and trade restrictions. While Western nations may be able to absorb the costs of such policies, people in the developing world will pay a higher price as many continue to suffer from starvation.

WHAT IS AGRICULTURAL BIOTECHNOLOGY?

Broadly speaking, biotechnology is any technique that uses living organisms or substances from these organisms to make or modify a product for a practical purpose. Biotechnology can be applied to all classes of organism-from viruses and bacteria to plants and animals-and it is becoming a major feature of modern medicine, agriculture and industry. Modern agricultural biotechnology includes a range of tools that scientists employ to understand and manipulate the genetic make-up of organisms for use in the production or processing of agricultural products.

Some applications of biotechnology, such as fermentation and brewing, have been used for millennia. Other applications are newer but also well established. For example, micro-organisms have been used for decades as living factories for the production of life-saving antibiotics including penicillin, from the fungus Penicillium, and streptomycin from the bacterium Streptomyces. Modern detergents rely on enzymes produced via biotechnology, hard cheese production largely relies on rennet produced by biotech yeast and human insulin for diabetics is now produced using biotechnology.

Biotechnology is being used to address problems in all areas of agricultural production and processing. This includes plant breeding to raise and stabilize yields; to improve resistance to pests, diseases and abiotic stresses such as drought and cold; and to enhance the nutritional content of foods. Biotechnology is being used to develop low-cost disease-free planting materials for crops such as cassava, banana and potato and is creating new tools for the diagnosis and treatment of plant and animal diseases and for the measurement and conservation of genetic resources. Biotechnology is being used to speed up breeding programmes for plants, livestock and fish and to extend the range of traits that can be addressed. Animal feeds and feeding practices are being changed by biotechnology to improve animal nutrition and to reduce environmental waste. Biotechnology is used in disease diagnostics and for the production of vaccines against animal diseases.

Clearly, biotechnology is more than genetic engineering. Indeed, some of the least controversial aspects of agricultural biotechnology are potentially the most powerful and the most beneficial for the poor. Genomics, for example, is revolutionizing our understanding of the ways genes, cells, organisms and ecosystems function and is opening new horizons for marker-assisted breeding and genetic resource management. At the same time, genetic engineering is a very powerful tool whose role should be carefully evaluated. It is important to understand how biotechnology-particularly genetic engineering-complements and extends other approaches if sensible decisions are to be made about its use.

UNDERSTANDING, CHARACTERIZING AND MANAGING GENETIC RESOURCES

Farmers and pastoralists have manipulated the genetic make-up of plants and animals since agriculture began more than 10 000 years ago. Farmers managed the process of domestication over millennia, through many cycles of selection of the best adapted individuals. This exploitation of the natural variation in biological organisms has given us the crops, plantation trees, farm animals and farmed fish of today, which often differ radically from their early ancestors.

The aim of modern breeders is the same as that of early farmers-to produce superior crops or animals. Conventional breeding, relying on the application of classic genetic principles based on the phenotype or physical characteristics of the organism concerned, has been very successful in introducing desirable traits into crop cultivars or livestock breeds from domesticated or wild relatives or mutants. In a conventional cross, whereby each parent donates half the genetic make-up of the progeny, undesirable traits may be passed on along with the desirable ones, and these undesirable traits may then have to be eliminated through successive generations of breeding. With each generation, the progeny must be tested for its growth characteristics as well as its nutritional and processing traits. Many generations may be required before the desired combination of traits is found, and time lags may be very long, especially for perennial crops such as trees and some species of livestock. Such phenotype-based selection is thus a slow, demanding process and is expensive in terms of both time and money. Biotechnology can make the application of conventional breeding methods more efficient.

GENOMICS

The most significant breakthroughs in agricultural biotechnology are coming from research into the structure of genomes and the genetic mechanisms behind economically important traits. The rapidly progressing discipline of genomics is providing information on the identity, location, impact and function of genes affecting such traits-knowledge that will increasingly drive the application of biotechnology in all agricultural sectors. Genomics sets the foundation for post-genomics activities, including new disciplines such as proteomics and metabolomics to generate knowledge on gene and protein structure, as well as their functions and interactions. These disciplines seek to understand systematically the molecular biology of organisms for their practical use.

A vast range of new and rapidly advancing technologies and equipment has also been developed to generate and process information about the structure and function of biological systems. The use and organization of this information is called bioinformatics. Advances in bioinformatics may allow the prediction of gene function from gene sequence data: from a listing of an organism's genes, it will become possible to build a theoretical framework of its biology. The comparison across organisms of physical and genetic maps and DNA sequences will significantly reduce the time needed to identify and select potentially useful genes.

Through the production of genetic maps that provide the precise location and sequences of genes, it is apparent that even distantly related genomes share common features. Comparative genomics assists in the understanding of many genomes based on the intensive study of just a few.

For instance, the rice genome sequence is useful for studying the genomes of other cereals with which it shares features according to its degree of relatedness, and the mouse and malaria genomes provide models for livestock and some of the diseases that affect them. There are now model species for most types of crops, livestock and diseases and knowledge of their genomes is accumulating rapidly.

AGRICULTURAL BIOTECHNOLOGY REGULATION

For traditionally bred plants, regulators rely on plant breeders to conduct appropriate safety testing and to be the first line of defence against genetic alterations that might prove dangerous. The same should be true for agricultural products developed using biotechnology. There is no reason for regulators to treat biotech plants any differently than traditionally bred plants, particularly given the fact that biotechnology provides greater control over gene manipulation. Despite fears about gene manipulation, traditional crossbreeding has altered plant genes and improved the human diet for the past 30 years. In fact, people worldwide safely consume these plants every day. Examples include wheat, corn, potatoes, tomatoes, and countless other staples of the American diet. In 1984, the biotechnology industry began experimenting with "gene-spliced plants" — the more advanced approach to gene manipulation that we now call "biotechnology."

DEPARTMENT OF AGRICULTURE

USDA regulates the release of all genetically engineered agricultural plants under statutes giving USDA's Animal and Plant Health Inspection Service (APHIS) the authority to regulate plants that may be or may become weeds or other nuisances — what the statutes call "plant pests." Although the rules apply in a general sense to novel or exotic varieties of both gene-spliced and conventional plants, APHIS typically only requires field testing of conventional plants that are new to a particular U.S. ecosystem (transplanted from another continent, for example).

However, all genetically engineered agricultural plants face a higher regulatory bar — making it more difficult to expand the use of biotechnology. Genetically engineered crops must be field tested under APHIS regulations prior to commercialization. For example, a new variety of corn produced with conventional hybridization requires no government-mandated field testing, but all new varieties of genetically engineered corn do, even though there is no logical reason for the regulatory disparity. For most genetically engineered plants, APHIS requires the company producing the plants to submit notice detailing the gene or genes that have been inserted, where the plant will be tested, and other relevant characteristics of the plant prior to receiving permission to conduct the field trials. Once the company completes field testing, APHIS reviews the results and makes a determination on whether or not the product should be "deregulated" and can be released into the market.

REGULATORY SCHEME

At the time, the White House Office of Science and Technology Policy began crafting a framework wherein existing federal agencies would regulate genetically engineered organisms on the basis of their characteristics, not the method of production — thus wisely deferring to the scientific community's judgment that regulation ought to address the products, not the process of biotechnology.

Engineered organisms would not require extra scrutiny simply because genetic engineering produced them (the process). Instead, they would be subject to heightened scrutiny only if the individual organisms expressed characteristics that posed some conceptually heightened risk (the product).

The federal government divided regulatory jurisdiction among agencies already involved in agricultural, food, and environmental regulation. These include the United States Department of Agriculture (USDA), the Environmental Protection Agency (EPA), and the Food and Drug

Administration (FDA). While each of these agencies considers the characteristics of individual products in their regulation, only FDA followed the general scientific thinking that genetically engineered and non-genetically engineered products should be regulated similarly. Both USDA and EPA automatically subject all genetically engineered plants as a class to premarket approval requirements not ordinarily applied to conventionally bred plants.

ENVIRONMENTAL PROTECTION AGENCY

EPA regulates plants that are modified to produce substances that act like pesticides: that is, substances used by a plant to protect itself from pests, such as insects, viruses, and fungi. EPA's proposed rule for these "plant-incorporated protectants" is not yet finalized. In the interim, however, EPA has regulated such plants by applying its proposed guidelines, which are functionally similar to rules for the registration of synthetic chemical pesticides. Again, biotech products face higher regulatory hurdles, even though plant-incorporated protectants developed through conventional breeding are exempt from these requirements.

HEALTH AND HUMAN SERVICES: FOOD AND DRUG ADMINISTRATION

FDA is responsible for ensuring that food items, including foods derived from genetically modified plants, are safe to eat. Following the general regulatory framework that emphasizes product regulation rather than process regulation, FDA rightly does not treat foods derived from genetically engineered plants to be inherently unsafe. Food producers are not required to seek premarket approval from FDA unless there is a substantive reason to believe that the novel trait (or traits) in the food pose a safety question. The initial determination of safety is left to the producer, but FDA has encouraged producers to consult with agency scientists prior to marketing a food produced with biotechnology to ensure that the appropriate determination is made. Recently, FDA published a proposed rule that would require producers to notify the agency at least 120 days before marketing, which may be a signal that FDA could abandon its more reasonable approach in the future.

In addition, FDA does not require labelling of foods derived from biotechnology unless the genetic insertions so alter the food that the common name no longer describes it adequately. Examples of this phenomenon would include such alterations of the food product that would raise a safety issue or change the product's nutritional composition or its storage or preparation characteristics.

INTERNATIONAL TRADE

While U.S. consumers do not appear to be strongly opposed to biotech foods (they, in fact, seem rather indifferent), a strong anti-biotechnology movement has arisen in several European countries over the last few years. The European Union (EU) has established strong restrictions on the commercial planting of genetically engineered plants, and European food processors and retailers are reluctant to import harvested agricultural products derived from biotechnology — thus jeopardizing the marketability of U.S. commodity grain exports. And the EU is now negotiating for strong restrictions on agricultural biotech products in important international agreements governing food safety and environmental protection. Very strong restrictions were included in the Biosafety Protocol, finalized in January 2000.

Perhaps, more important, though, are the Codex Alimentarius Commission standards for food safety. Without convincing scientific evidence that genetically engineered crop plants pose a heightened environmental or human health risk, restrictions on agricultural biotech imports could be challenged under the General Agreement on Tariffs and Trade and the World Trade Organization.

Thus the EU has resorted to justifying its decisions on the basis of a risk-management philosophy known as the "precautionary principle," and the EU advocates inclusion of the precautionary principle in international environmental and food safety agreements (such as the Codex Alimentarius), as well as within GATT itself. Its inclusion would give nations greater latitude in restricting imports of U.S. agricultural products.

No single definition exists for the precautionary principle, but its general meaning is that when an activity raises threats of harm to human health or the environment, regulatory measures should be taken to prevent or restrict the activity even if the risks have not been demonstrated scientifically. Although the EU asserts that the precautionary principle is an unbiased risk-management philosophy, critics have noted that its lack of definition and evidentiary standards makes it all too easy to abuse for the purpose of masking trade protectionism, and that its very approach to risk management is inherently flawed and may, in fact, increase net risk.. Nevertheless, the precautionary principle has already been included in several international environmental treaties, including the Convention on Biological Diversity and the Biosafety Protocol.

LABELLING ISSUES

Some activists, however, argue that the government should mandate the labelling of all genetically engineered foods. They assert that consumers have a "right to know" how their foods have been altered, and that a mandatory label would best allow consumers to choose between genetically engineered and conventional foods. Biotechnology advocates have argued against mandatory labelling because such requirements raise food costs— something that mostly harms lower-income Americans and people on fixed budgets. Perhaps more important, while biotech products are not substantially different from other products, special labels would likely make consumers think these products were more dangerous. Hence, rather than serving educational or "right to know" purposes, such labels promise to simply confuse consumers.

A government-mandated label on all genetically engineered foods also would raise important First Amendment free speech issues. In 1996, the U.S. Court of Appeals, in *International Dairy Foods Association et al. v. Amestoy*, ruled unconstitutional a Vermont statute requiring the labelling of dairy products derived from cows treated with a genetically engineered growth hormone, noting that food labelling cannot be mandated simply because some people would like to have the information. "Absent ... some indication that this information bears on a reasonable concern for human health or safety or some other sufficiently substantial governmental concern, the manufacturers cannot be compelled to disclose it."

In other words, to be constitutional, labelling mandates must be based in science and confined to requiring disclosure of information that is relevant to health or nutrition. Furthermore, consumers need not rely on mandatory labelling of biotech foods to truly have a choice. Real-world examples show that market forces are fully capable of supplying information about process attributes (including kosher and organic production standards) that consumers truly demand.

The same can be said about nonbiotech foods, and the FDA recently published proposed guidelines to assist producers in voluntarily labelling both genetically engineered and nongenetically engineered foods. Additionally, the USDA's newly published rule for organic certification necessarily excludes biotech products from organic food production.

Consequently, consumers wishing to purchase nonbiotech foods need only look for certified organic products.

Even as underdeveloped nations clamor for biotechnology applications, and as countries like China continue to experiment with and use agricultural biotechnology, opponents of agricultural biotechnology in the West, particularly Europe, attack it as an unnatural process that will destroy the world, not better it. They argue that biotechnology should be heavily regulated, if not banned. Already, however, genetically engineered plants are subject to strict regulatory oversight that is equal to or greater than that advocated by the vast majority of scientific specialists.

Additional regulation will slow down research and development of genetically engineered crops, keep beneficial products off the market, and raise the cost of products that do make it to consumers. Furthermore, the inclusion of similar restrictions — or inclusion of the precautionary principle— in international agreements will greatly impact the international trade of agricultural goods and delay their introduction into the marketplace. Each of these problems could prevent this technology's benefits from being introduced to industrialized nations and, more important, the developing world.

AGRICULTURAL BIOTECHNOLOGY SECTOR IN INDIA

Agricultural biotechnology sector in India, particularly its plant biotechnology segment, is uniquely poised for a major process of transformation. As the country's agricultural biotechnology sector made strides in the 1990s in r-DNA, transgenics and molecular marker assisted plant breeding process, the Government of India responded with a matching policy support and regulatory framework that was designed to render the path of progress in R&D, sustainable and bio-safe. To a large extent, developments in the policy front have been induced by a vibrant non-governmental sector that intensely intervened on the sensitivities of modern biotechnology. This chapter discusses analysis of the issues and proceeds to further consider the path of progress achieved by agricultural biotechnology companies in India in the area of commercialization of biotechnology products. The special focus of this chapter is on plant biotechnology. The biotechnology sector in India in 2002-03 is estimated to have attracted private investments to the tune of US$ 140 million. Infrastructure and R&D account for nearly 80 per cent of these investments.

During the period from 1986 to 2002 it is estimated that the Government of India invested an amount of US$ 275 million in the biotechnology sector. The Government of India has also extended venture capital support for biotechnology start-ups through the Technology Development Board and the New Millennium Indian Technology Industry Leadership Initiative (NMITLI). Meanwhile the States of Andhra Pradesh, Karnataka, Maharashtra, Gujarat, Kerala and Tamil Nadu have drawn up ambitious sub-national biotechnology policies, which promise fiscal and infrastructure support measures to prospective entrepreneurs. Some State Governments have launched their own venture funds for supporting novel start-up ventures.

To catalyse the process, the Government of India accorded priority to IPR protection. The Indian Patents Act of 1970 was amended to provide for product patents for agro-chemicals, drugs and pharmaceuticals and micro-organisms over a transitory period. A product patent regime would be in place in India by January 1, 2005. The new patent law extends protection term for inventions from 14 to 20 years. To provide for protection of IPRs in respect of new plant varieties, Parliament has enacted a *sui-generis* legislation in the shape of the Plant Varieties Protection and Farmers Rights (PVFR) Act in 2001. These developments have created favourable legal conditions for international partnerships in biotechnology R&D. By prioritizing FDI investments in biotechnology, the Government of India has sent appropriate signals to international investors. Further the fact that these changes have been in consonance with the WTO-TRIPs has lent greater creditability to these changes. Meanwhile, the Government has also brought in regulations over agricultural biotechnology. Apart from the biosafety regulations that form part of the

Environment (Protection) Act 1986, the Indian Parliament has legislated the National Biodiversity Act in the year 2002, to provide for regulatory controls over access to biological resources in India.

AGRICULTURAL BIOTECHNOLOGY INDUSTRY

The top ten pesticide and seed firms for 1997 and 1999. Longer-term time series data on the structure of the chemical and seed company are difficult to obtain. In one of the few published papers to provide this data, Ollinger and Fernandez-Cornejo (1998) provide Concentration Ratio data (CR4) for the United States pesticide industry over the period 1972-1989. During this period, the CR4 oscillates, moving from a high of 50% in 1973 to a low of 37% in 1982, and then rising to 48% in 1989. Although detailed data are not available for the 1990s, there is evidence that the structure of the seed and chemical industry remains dynamic.

The consolidation activity of the ten most active biotechnology firms. Mergers and acquisitions are a key component of the new business relationships that have emerged. As Kalaitzandonakes and Hayenga (2000) note, firm entry in the crop biotechnology industry peaked in the early 1980s, with production innovation continuing throughout the 1980s. Oehmke et al. (1999) also discuss the cyclical pattern in mergers in the biotechnology industry. Aggregate market figures mask the much higher concentration that exists in specific markets. For example, in 1998, Monsanto and Pioneer-HiBred (now owned wholly by DuPont) controlled 15% and 39% of the US seed corn market, respectively. These two same companies controlled approximately 24% and 17%, respectively, of the purchased soybean seed market. For the US cotton market, two companies, Delta & Pine Land and Stoneville, had 71% and 16%, respectively, of the seed market.

The determination of the relevant market concentration is not always done on the basis of output markets. The Federal Trade Commission has used innovation competition to assess the impact of mergers. Examining competition in innovation is designed to focus attention on the impact of a merger on innovative activity. Using data on field trials underway each year, Brennan Pray, and Courtmanche (2000) calculate a CR4 ratio at the R&D stage. In 1988, the four largest firms had 87% of the field trials. The CR4 ratio declined to a low of 63% in 1995, then rose steadily over the next few years to reach 79% in 1998. This concentration at the R&D stage is matched by concentration in terms of the number of patents held. The top four firms held 41% of the corn patents (up to 1996), 53% of the soybean patents (up to 1997), 77% of the tomato patents (up to June 1997) and 38% of the Bt patents (up to 1998).

Concentrated markets do not necessarily imply the presence of market power. Baumol, Panzar and Willig (1992) stress that firms will not be able to exercise market power if the market is contestable. The key requirements for market contestability are: (a) Potential entrants must not be at a cost disadvantage to existing firms, and (b) entry and exit must be costless. For entry and exit to be costless, there must be no sunk costs. Sunk costs are expenditures that cannot be recouped once they are incurred—examples include expenditures made to obtain regulatory approval, expenditures on advertising, and expenditures on R&D. If there are no sunk costs, potential entrants can use a hit-and-run strategy in which they enter an industry, undercut the price of the incumbents, reap the profits and exit before the incumbents have time to retaliate. In anticipation of entrants acting in this manner, the incumbents forestall entry by keeping price at average cost. The consequence is that, even in an industry that is highly concentrated, prices can be kept at or near competitive levels.

However, if sunk costs are present, firms entering an industry are unable to exit again without losing a portion of their investment. As a result, hit-and-run strategies are much less profitable and incumbents are able to keep price above average cost. Thus, with sunk costs, markets are not contestable and market power is once again an issue.

VERTICAL STRUCTURE

In addition to becoming larger through growth and horizontal integration, the major agrochemical companies have restructured themselves in other ways. One of the major changes is increased vertical integration—for example, the combining of seed, chemical and biotechnology activities within the same firm. Although some companies initially tried to include pharmaceutical and biotechnology components within the life science model, practically all of them have now divested their agricultural operations. A further element of this internal restructuring is the increased use of strategic alliances.

Many countries have had their own local seed companies that have over the years developed seed for a specific geographical market and operated sales and distribution systems. The major biotechnology companies are increasingly purchasing these seed companies as a source of seed material in which to insert the genes for herbicide or insect resistance. As an example, in 1997, Monsanto acquired a 30% share of the Brazilian corn seed market with the acquisition of Sementes Agroceres.

With its 1998 purchase of Cargill's international seed division, Monsanto now controls over half the Argentine maize seed market. In 1998, Dow AgroSciences acquired Morgan Seeds, Argentina's second largest corn seed company, and Brazil's Dinamilho Carol Productos Agricolas, another key South American corn seed company. Phytogen (majority owned by Dow Agrosciences) acquired a major cottonseed breeding programme in the Chaco province of Argentina. In 1998, Mexico-based Empresas La Moderna (ELM) bought two South Korean vegetable seed companies and Nath Sluis (agricultural biotech company) of India (Action Group on Erosion, Technology and Concentration [ETC], 2000).

FACTORS THAT AFFECT HORIZONTAL INDUSTRY STRUCTURE

Numerous factors are behind the wave of horizontal mergers and acquisitions. Some of these factors are common to all industries and have no specific link to the seed, chemical and agricultural biotechnology industries. Thus, the mergers and acquisitions in the seed and chemical industries are at least in part a result of the need to consolidate costs and rationalize industry capacity, a desire by the management of the firms involved to extend their sphere of influence, and a wish by some firms to pre-empt other firms from taking over valuable assets. MacDonald (2001) provides an overview of concentration levels in a number of US agricultural sectors, many of which have seen higher concentration in recent years.

There are at least two explanations for increasing concentration, however, that pertain specifically to the seed, chemical and agricultural biotechnology industries: (a) sunk costs, and (b) escalation strategies.

SUNK COSTS

Two sources of sunk costs appear to be important in determining industry concentration: (a) intellectual property rights and R&D expenditures, and (b) the regulatory requirements that governments have introduced before the products of the R&D activity can be marketed.

On the empirical side, Ollinger and Fernandez-Cornejo (1998) examine sunk costs and regulation in the US pesticide industry. Using data over the 1972-89 period, they find that research costs and pesticide regulation costs negatively affect the number of companies in the industry, and that smaller firms are affected more strongly by these costs than are larger firms. Research and regulation costs also encourage foreign-based firms to expand into the US market and to force less profitable innovative firms to exit the market. Ollinger and Fernandez-Cornejo also point out that their results on the impact of regulatory costs generally match those found in other industries.

These empirical results can be linked to theoretical arguments concerning sunk costs and the economies of scale and scope that sunk costs tend to create. As Sutton (1991) outlines, the presence of sunk costs means that for firms to be profitable, price needs to be raised above marginal cost, typically by reducing the amount of competition (*i.e.*, the number of firms). Sutton identifies two types of sunk costs: exogenous and endogenous. Exogenous sunk costs are those beyond the control of the firms in the industry. Regulatory costs are a good example of exogenous sunk costs. In contrast, endogenous sunk costs are firm-level strategic variables. Advertising and R&D are examples of endogenous sunk costs. The rest of this section examines the connections between intellectual property rights (IPRs), R&D expenditures, regulatory costs, and economies of scale and scope.

Economies of scale and scope are major factors in propelling industries towards concentration. Economies of scale exist when average costs fall as more output is produced; economies of scope exist when the total cost of producing two outputs together is less than the cost of producing the two outputs separately.

Because economies of scale and scope mean that larger and diversified firms have lower average costs, there is clearly an incentive for firms to get large. Indeed, those that do not get large are vulnerable to being driven out of the market by larger and more cost efficient firms. Of course, there is a limit to how large firms can get. Although development and production costs may fall with an increase in the size of the firm, other costs—particularly those associated with administration—rise. Nevertheless, economies of scale and scope clearly create pressures for consolidation.

Economies of scale and scope are created as a result of investment in non-rival goods. Unlike rival goods—such as materials, labour, and energy—which can only be used in one place, by one person and at one time, non-rival goods can be used in more than one place and by more than one person, all at the same time. This feature of non-rival goods—that they can be used over and over again—means that output can be increased without having to increase all inputs. Consequently, economies of scale and scope are created. Much of the expenditure on non-rival goods—examples include R&D and regulation expenditures—can be considered a sunk cost.

Intellectual property, of course, is a good example of a non-rival good. Indeed, ideas generally are considered non-rival goods. It is widely believed that many of the high-technology industries, including the biotechnology and information industries, are subject to increasing returns. Romer (1990) postulates that this is a result of the distinction between physical goods and ideas. Ideas are not scarce, and as such, any industry based primarily on the trade of intellectual property will not face diminishing returns in their primary resource, the idea. The protection of intellectual property, through IPRs, is designed to encourage innovation and to provide incentives for the development and diffusion of new products and technologies.

An example illustrates the connection between intellectual property and economies of scale and scope. Suppose a biotechnology firm has some intellectual property, such as a particular gene that has been isolated. This intellectual property can be used over and over again as the firm expands its activities. For instance, if the company wishes to develop seeds for a new crop, it will not have to invest again in the research that isolated the gene. Although the development of a new seed will require additional lab and greenhouse space, labour, and materials, the expenditure on the technological advancements does not have to be made again.

A similar result occurs if the firm needs to invest substantial amounts of money in obtaining regulatory approval for a seed—although the production of additional units of the seed will require additional costs, the regulatory expenditures do not have to be made again. Once again, large companies typically have an advantage, because they are able to spread the costs of obtaining regulatory approval over more output. Thus, the greater the regulatory requirements in an industry, the more concentrated the industry is expected to be.

Intellectual property may also create economies of scope. Once a specific gene has been isolated—for instance, a gene that confers a resistance to a particular herbicide—this gene can then be put in a number of crops. Once again, the production costs of a number of products together will be less than if the products were produced separately.

To recap, R&D expenditures and regulatory costs are both sunk costs and a source of economies of scale and scope. Because economies of scale or scope mean that larger and more diversified firms have lower average costs, there is clearly an incentive for firms to get large. Indeed, those that do not get large are vulnerable to being driven out of the market by larger and more cost efficient firms. These theoretical results are supported by empirical observations in the US pesticide industry.

ESCALATION STRATEGIES

An escalation strategy is one in which a company spends large amounts on R&D and engages in mergers and acquisitions to achieve a dominant role in the market—that is, the firm tries to leapfrog its competitors to become the dominant firm. Escalation can be a profitable strategy when there is a high degree of substitutability with competitors' products on the demand side, and there are scope economies on the supply side. Both of these factors are present in the agricultural biotechnology industry.

The isolation of a gene that provides particular advantages and which can be inserted into a number of crops means there are scope economies on the supply side. There are also clear supply side scope economies associated with the enabling technologies that are required to use these genes. And on the demand side, herbicide and insect resistant seeds and the accompanying chemicals are clearly a substitute product for traditional seeds and herbicides and pesticides.

Firms engage in escalation strategies in a number of ways. Although internal growth through R&D is one way, the more common method is through consolidation via horizontal mergers and acquisitions. As the theory predicts, the combination of demand substitutability and supply side scope economies present in the seed and chemical industry appears to be linked with escalation strategies.

Monsanto has clearly followed an escalation strategy, with Dow, Novartis, and others following similar strategies of making acquisitions and spending significant amounts on R&D.

IPRS AND INVESTMENTS IN PLANT BIOTECHNOLOGY

The implicit rationale for IPR protection is that they promote investments in plant breeding and bio-engineering. This hypothesis is contested by section of analysts. Nevertheless it is a fact that the structure of IPRs do influence investments. IPRs induce their own pattern of innovations. The time taken for R&D to fructify as innovations is a crucial aspect guiding investments.

Often IPR regimes play a key role in influencing the time pattern of innovations. It is a well-known fact that conventional plant breeding methods are dilatory and time consuming. It requires 7 to 8 generations of repeated breeding to reduce heterozygosity of new genotypes. As Vossen (1985) observes Arabica Coffee subject to three to four cycles of breeding still display heterozygosity and it is not until 20 to 25 years (spanning 4 to 5 generations) that one arrives at true breeding varieties. This time lag is even more pronounced for the Arabusta coffee variety (inter-specific hybrid of Arabica and Robusta Coffee) where segregation of unfavourable characteristics could still occur even after several generations of backcrossing and selfing.

Another interesting example is the one cited by Evenson (1991) about the long gestation period in regard to rice varieties in India. According to Evenson, India released 306 rice varieties for planting during the period 1965-1986 after performing 20,000 crosses over a period of over 15 years since 1950. It must be understood that this long process of backcrossing was preceded by considerable time spent on location and identification of individuals plants with requisite traits. The speed or pace at which plant varieties qualify for being released for planting have economic implications on investment behaviour in seed/plant breeding industry. Asexual modes of plant breeding which include plant genetic 'engineering' and transgenic techniques could obviate the long gestation period associated with pre-release preparations encountered conventional breeding.

It is a different matter than biosafety concerns may delay releases of these varieties. The time lag in release of plant varieties fulfilling the criteria of novelty, distinctness, uniformity and stability (or alternatively the requirements of novelty, inventiveness and commercial application) raise fundamental issues from the view-point of capital investments. The time lag in varietal release is potentially least in the case of transgenic and non-sexually produced varieties while for traditional or classical methods the time lag can be greater.

The UPOV 1991 based plant IPR regimes could induce longer time lags as compared to the situation when an IPR regime is based on the 1978 version of the UPOV. This is due to the fact that the minimum genetic distance concept implicit in UPOV 1991 precludes 'close cousin' varieties from acquiring protection rights, unless the techniques of genetic distance determination by themselves are inadequate. UPOV 1978 would facilitate capital investments to some extent as the release of plant varieties is periodically more frequent. Even under the plant patent regimes, investments could bring about short-term returns, though the volume of investments will have to be larger on account of capital intensity of advanced plant genetic engineering technologies. By comparison, UPOV 1991 regime is least conducive to capital investments as plant variety releases can be painfully slow under this regime. Therefore sustenance of conventional plant breeding under UPOV 1991 regimes, can only be ensured by infusion of low discount capital.

Paradoxically it is the reduced availability of low discount public funds for plant breeding that has induced countries like India to go in for Plant Varieties Protection Legislations modelled on UPOV lines. The implicit assumption in undertaking this step is that private funds for R&D in plant breeding will 'crowd in', if protection plant varieties rights are accorded. But the reality is that in liberalised deregulated developing economies, private investments have 'crowded in', in short return promising avenues.

Consequently classical plant breeding activities will suffer for want of funds in the UPOV 1991. In other words, UPOV 1991 modelled national legislations will not necessarily hold attraction for private sector investors in R&D projects involving classical plant breeding as compared to say, UPOV 1978 or the Plant Patent regimes. Rather private investments would be catalysed better by *sui-generis* plant protection legislations, which partake of plant patent elements and focus in the main on hybrids and genetically engineered varieties.

Given the linkages between plant breeding and the seed industry these perverse investment behaviour trends can produce major economic consequences. The Seed Laws in various countries, both advanced and developing, are getting to be progressively oriented towards restricting market circulation to varieties that are conferred with plant breeder or patent rights. The UPOV guidelines on DUS are used world wide not only as the basis for establishing varietal distinctiveness and descriptions but also for seed certification purposes. The DUS guidelines are employed for recognising and registering not only the 'basic' seeds used for multiplication of 'pedigree' seeds but also for certification of the pedigree seeds themselves.

IPR REGIMES FOR PLANT BIOTECHNOLOGY

Research and development in the agricultural sector is unique among industries in at least two respects: the truly global reach of a majority of agricultural R&D and the historical success of what has been largely a public enterprise. In relation to other industries, research and innovation in agriculture are far more geographically dispersed. Private sector ?rms make up roughly one-third of global agricultural R&D expenditures, while public research institutions make up the other twothirds, evenly split between developed and less developed countries. Intellectual property rights (IPR) are important internationally, in public, nonpro?t and private institutions, both as incentives for innovation and as constraints on freedom to operate in agricultural R&D.

Article 27(3) of the TRIPs Agreement lays down that members shall provide for the protection of plant varieties either by patents or by an effective *sui-generis* system or by any combination thereof. Article 27(3) in particular is critical to India as this requires us to provide protection of plant varieties by conferring plant breeders rights either by way of patents or through *sui-generis* plant variety protection legislations. Countries with strong R&D base in plant genetic engineering such as USA have robust Plant Utility Patent Legislations. India is certainly not inclined to adopt patent protection regimes for its plant varieties. India is rather inclined to adopt a *sui-generis* legislation which is non-'patent' based. There are many reasons for this proclivity. India is one of the ten mega-diversity countries of the world and a rich storehouse of landraces of principal agricultural crops. India has a strong R&D base in conventional plant breeding methods. Its strength in plant genetic engineering is impressive but not an overwhelming factor by comparison. The first two strengths explain India's disinclination towards plant patent regimes or towards a *sui-generis* legislation which is 'patent'-driven.

While Plant Utility Patents Act provide for broad patents over plant varieties, traits and genes and even the physical parts of the plants, plant breeders rights provide for IPR only over varieties. As is well known, since 1990s the UPOV (Union for the Protection of Varieties) has been largely viewed by developing countries as offering the best regime for positioning their national legislations.

The central feature of the UPOV is the protection it affords to plant breeders who produce plant varieties that fulfil the criteria of distinctiveness uniformity and stability (DUS). The current version of the UPOV, *viz.* UPOV 1991 had added additional criteria of 'new' to DUS thus rendering DUS as NDUS. Contrary to the popular notion, the NDUS criteria of UPOV 1991 is not substantively different from the principles of 'novelty', 'inventiveness' and 'industrial application' (NII) which applies for patents. The criteria of 'novelty' and 'inventiveness' in Patent Laws are covered by the criteria of 'new' and 'distinct' in UPOV 1991. Thus by distinctness, the UPOV means a variety of plant which is 'clearly distinguishable from other varieties whose existence is a matter of common knowledge'. It is apparent that this term captures the attributes of 'novelty' and 'inventiveness' implicit in Patent Laws. Even in respect of 'uniformity' and 'stability' criteria the UPOV does not offer different recipes. True, by 'stability', the UPOV conveys that 'relevant characteristics of protected plant variety remain unchanged either for a specified period or after repeated propagations or cycles of propagations'. It is also true that 'stability' is a difficult criteria for a plant breeder to fulfil.

Attainment of 'stability' criterion is problematic for cross-pollinated plants and non-single homozygous lines of autogamous plant varieties. This, in turn, reduces the commercialisation potential of the plant variety since inconsistency of genetic quality jeopardises commercial application of the plant variety. The same holds true of the criterion of "uniformity". Therefore, the NDUS criteria of UPOV 1991 are homologous to the NII criteria implicit in Plant Utility Patent Laws.

The UPOV has undergone two major amendments since its inception in 1961, *i.e.* in 1978 and 1991. The main concern with regard to currently operational UPOV 1991 has been its stringent provisions, which include the additional attribute of 'new' to the DUS criteria. The scope of UPOV 1991 in terms of coverage is larger. While plant varieties of all taxa are covered by UPOV 1991 only plant varieties of nationally designated species were covered by UPOV 1961 and 1978. Further by defining clearly that 'variety' represents a 'group of plants', or 'single' or 'several plants' or 'one' or 'several parts' of a plant(s) the UPOV 1991 has considerably enlarged the scope of its coverage. Going by this definition even sexual and vegetatively propagated varieties are covered by UPOV 1991. The US PVPA legislation has been quick to adopt the UPOV 1991 formula. The amendments of 1994 to the US, PVPA broadens the term 'plant varieties' to include all materials harvested from the protected varieties. The other drastic feature of UPOV 1991 has been abridgement of the 'plant breeders exemptions' and 'farmers privileges' conferred by the earlier versions.

Earlier versions of UPOV allowed farmers to re-use seeds generated from their farms for self-use, and allowed breeders to freely use protected varieties for further improvements. These concessions gave UPOV a 'less stringent' face in comparison to 'patent' laws. UPOV 1991 gives an option to National Governments to disallow the farmers privileges of retaining or re-using seeds for self cultivation. UPOV 1991 also restricts the breeders rights by enlarging the right of the breeder to close variations of his cultivar or 'essentially derived varieties'. In other words, the breeder of a cosmetically bred variety would have to buy genetic dependency rights from the derived variety prior to commercialising the derived version. Further the duration of protection have also been extended under UPOV 1991 to 20 years for crops and to 25 years for trees and vines. Thus in terms of these rigorous features, UPOV 1991 eliminates the arbitrage between plant varieties protection and patent regimes.

NEW TRENDS IN AGRICULTURE

Since time began we humans have always relied on plants and animals to provide us with food, shelter, clothing and fuel, and for thousands of years farmers have strived to better these in order to continue providing us and to meet the needs of evolution. As the population grows and expands then so does the need for resources provided by plants and animals. With the population expected to reach over 10 billion by the year 2030 it is estimated that the world food production will have to double on the farmland that exists today, if it is to compete with the anticipated growth in the population. The ever-increasing needs of yields of crops and the decreasing crop inputs such as water and fertiliser can all be helped by biotechnology as can advancements in providing pest control that is better for our environment.

POST R&D REGULATIONS AND INVESTMENTS

Post-R&D regulations play a critical role in deciding on the pace of investments in the biotechnology sector. Even if a start-up enterprise enjoys a favourable IPR regime, that rewards its inventions, it is unlikely that it will invest in product development, in case post-R&D regulations are rigid and dilatory in nature. This then forms the backdrop for strategic alliances between start-up ventures and downstream industrial complexes. Multi-layered regulations in the post-R&D phase involve complexities which are expressed in the delays in commercial application /utilization of invented products or technologies.

In the case of third generation plant biotechnologies, such as transgenics, the problem is compounded by the fact that regulations by themselves are complex and less understood by the regulators themselves.

A transgenic plant would have to undergo multi-stage regulatory checks and clearances before finding commercial application.

India is uniquely situated as far as biotechnology investments and infrastructure are concerned. In the area of plant biotechnology, advances have been made in transgenics of rice and wheat carrying stress tolerant genes such as CodA, COR47 and HVA1. The larger challenge is to take these products through the regulatory processes. More formidable is the issue of market development and appropriate pricing of transgenic seeds. There are many imponderables here. Market shares for seeds in India have varied from crop to crop and from region to region. The 'demand' for transgenic plants or seeds may be even more uncertain, given the 'safety' dimensions and risk aversion propensity of Indian farmers. In the dry agricultural pockets of Central and South India, farmers tend to be attached to traditional drought-hardy varieties of plants, which in turn contributes to their initial 'inertia' to shift to transgenics.

Thus the task of marketing and pricing of novel biotechnology plants gets to be complex in such environment of 'demand incoherence'. A high margin pricing strategy for seeds may choke demand for transgenics. Thus the degree of risks in commercializing transgenic plants could be of a high order. Given these realities, the idea of strategic alliances between start-ups and downstream entities looks to be realistic. The fact that premier R&D institutions such as the Indian Institute of Science, has set up its own start-up biotechnology venture, renders a strategy of tie-ups, a logical option. Much will depend upon the complexities inherent in product development and the willingness of competent downstream entities to signup agreements. To a large extent, this will depend upon enabling policies at the upper end of the value chain.

To sum up, India's agricultural biotechnology industry has the necessary policy and regulatory support to promote active R&D. However, in the critical area of product development and commercialization much more attention needs to be paid. The multi-tiered regulatory framework that exists could seriously affect the performance of start-up companies that would like to run through the entire life-cycle of biotechnology product development.

Consequently, partnerships and strategic alliances should be encouraged between start-up ventures and established companies that have high product development capabilities. The experience of other countries in the Asia Pacific region could be a key input in framing development strategies for biotechnology products in India.

BIOTECHNOLOGY INNOVATION SYSTEM IN INDIA

The government plays an important role in building the National Innovation System (NIS) by setting up institutions (public research institutions such as universities, labs etc), developing policies that support innovation and providing regulatory frameworks that can have a considerable influence on sectoral development.

In the context of the developing countries, the role of the government becomes more prominent as the majority of the resources for the scientific research and development are being channelised or governed by the government. In case of India, nearly 80% of the gross expenditure for research and development is being incurred by the Government while in case of the industrialized nations such as Japan, USA and UK the Industry spends nearly 75%, 64% and 44% respectively. Though there has been a significant increase in the country's Gross National R&D expenditure in absolute terms from Rs.3974 Crore in 1990-91 to Rs.16199 Crore in 2000-01 to Rs.18000 Crore in 2002-03, the share of the government in national R&D expenditure has remained unchanged. However, it is interesting to note that internationally all the governments perform the in-house R&D function, including even the highly private-sector-oriented governments such as the US and other OECD countries. A study of the Biotechnology innovation system of the four global leading countries namely USA, UK and Japan and Germany further supports the active involvement of Government R&D in furthering the growth of the Biotechnology sectoral innovation. Biotechnology is being characterised by multidisciplinary fields, high dependence on science and are emerging from industries having large R&D base, high degree of qualified manpower and having close links or interaction with public research institutes as well as universities and thus faces a broad set of different policy interventions by the government underlying the National Innovation System (NIS).

The National System of Innovation (NSI) perspective is a convenient framework to understand the process of innovation occurring in an economy and especially within the manufacturing sector. It focuses on the country-specific factors influencing the process of innovation and economic growth. The NSI approach has evolved over the years both analytically and empirically with the contributions from Freeman, Lundvall and Nelson, Edquist (1997, 2002). NSI has became popular as an analytical framework to study innovation at the national, regional and even Sectoral level.

Though there are several definitions of NSI but that best fit in to the present context is that of Metcalfe (1995). NSI is defined by Metcalfe (1995) as a "set of distinct institutions which jointly and individually contribute to the development and diffusion of new technologies and which provides the framework within which governments form and implement policies to influence the innovation process. As such it is a system of interconnected institutions to create, store and transfer the knowledge, skills and artefacts which define new technologies".

NSI framework is comprised of various components such as the government, independent research institutes, firms and higher education system etc and their interrelationships. Innovation or government policy instruments and institutions play a very important role in cementing the relationship among the various components for the success of NSI. An application of this framework shall help in analysing the systemic failure that hamper the generation of innovations. Public policy can be applied to correct such failure. There are several NSI studies on Biotechnology in the context of the developed world but the studies in the developing countries are rather limited.

Biotechnology NSI in India has evolved over time and has been instrumental in creating infrastructure and capabilities in promoting biotechnology innovation in the country. There are 241 institutions and 321 firms engaged in various dimension of biotechnology in the country. Indian Biotechnology industry is now among the top 12 countries in the world and like its global counterpart, is dominated by the healthcare sector. The recently released "Draft National Biotechnology Strategy 2005" by the Indian Government is a step forward in the direction of promotion and development of biotechnology to withstand global challenges and yet, solve the local healthcare needs of the country. Studies on biotechnology in India by and a recent empirical analysis of the health innovation system of seven developing countries further supports the role of the Governments in providing the sustainable environment for the successful development of health biotechnologies.

According to Acharya *et. al.* government in developing countries is one of the key actors in prioritizing, implementing new technologies and promoting innovation to address their specific health and development needs. The governments of the countries where biotechnology innovation system has been successful in perform or support the following functions/activities such as knowledge support, commercialization of research, collaborative research, financial support, regulation, IPR and resolving ethical issues etc.

Thus it would be interesting to analyse the role of the government in the promotion of the biotechnology in the country with special focus on the healthcare sector.

Hence, the present study traces the role of government or national approaches in terms of policies, institutions and instruments in the promotion of biotechnology innovation system in India with a special focus on the healthcare biotechnology sector. A systemic framework similar to the EPHOITE (2003) project is being adapted to examine the direction of public policies, institutions and instruments in addressing the issues of knowledge support, commercialization of research, collaborative research, financial support, regulation, IPR and ethical issues in the area of healthcare biotechnology in Indian context.

The significant features of national biotechnology innovation system with focus on healthcare biotechnology with the help of various S&T indicators. The various government initiatives in terms of various policies, institutions and instruments in the framework of NSI while the lessons learnt along with the policy implications and challenges for the growth of healthcare biotechnology innovation.

MODERN AGRICULTURAL BIOTECHNOLOGY

In 1973, Cohen and Boyer transferred a gene from one organism into another. In 1982, the first biotech plant, an antibiotic resistant tobacco, was developed. In January 1983, at a meeting of genetic researchers in Miami, three different teams reported success in using Agrobacterium tumefaciens, a bacterium, to carry new genes into plant cells, heralding the dawn of modern agricultural biotechnology. Agrobacterium tumefaciens, described as a "natural genetic engineer", splices its own genes into host plant cells. This pathogenic bacterium was now converted into a pack mule, to carry new, foreign genes into plant cells, minus the disease and this became the most common means of producing Genetically Engineered Organisms (GEOs).

Field tests for GE crops resistant to pests and pathogens were first conducted in the US, in 1985. A co-coordinated framework for the regulation of GEOs was established and the first GE tobacco was released, in 1986 . The US Department of Agriculture published guidelines for field trials of GE crops in 1991.

On approval from the Food and Drug Administration, Flavor Saver, the first GE tomato, with a longer shelf life, was on the US markets in 1994. During 1995-96, GE soybean, corn and cotton were approved for commercialization, in the US.

A number of GE crops, developed for pest, pathogen and herbicide resistance, are now commercially cultivated in several countries. Rice with pro-vitamin A, higher iron content, or human milk proteins and potatoes with high protein content are in various stages of development. Tobacco plants producing functional human haemoglobin and bacteria that produce human insulin have been developed, as well as plants with vaccines against rabies and other viral diseases. Food grain crops that withstand drought and salinity are high priority research, and so are those for high yield. A GE tobacco plant detoxifies soils contaminated by explosive residues, providing solution to a frustrating environmental problem in countries ravaged by armed conflict.

Now more than 70 biotech agricultural crops that have been approved for use in North America, including varieties of soybeans, cotton, canola, corn, potatoes, squash, tomatoes and papaya. About six million farmers in some 17 countries now cultivate GE crop on about 125 million acres, a 30 fold increase over 1996. By end of the year 2002, six GE crops planted in the US (soybeans, corn, cotton, papaya, squash and canola) produced an additional four billion pounds of food and fibre on the same acreage, improved farm income by US $ 1.5 billion and reduced pesticide use by 46 million pounds. In 2003 in the US, 80 per cent of soybean acres will be planted with biotech varieties.

A number of activist groups vehemently attack GE technology and its products on grounds of safety to humans and the environment, and costs of technology transfer and its reach to the needy. Products of agricultural biotechnology bear the brunt of this ill-informed, unscientific and prejudiced onslaught much more than GE products related to health or industry.

In 2001, the European Community released results of a 15-year study, costing US $ 64 million, and involving more than 400 research teams and 81 projects. This report concluded that GE products pose no more risk to human health or the environment than conventional crops. So far, extensive and intensive research on the probable risks of GE technology has not brought out any adverse effects and none of the fears expressed by anti-tech activists were proved even marginally. Nevertheless, caution and examining issues case by case, is the watch word of technologists, who are aware of their responsibilities.

For a number of products, technology transfer is free of costs for developing countries, as for example Golden Rice, the rice with pro-vitamin A. It is the responsibility of the Governments of the respective countries to bear the costs of developing local varieties and to reach the products to the needy at an affordable cost. In India, three GE varieties of pest resistant cotton were approved for commercialization, a year ago. Approval for other GE varieties of cotton and GE mustard was deferred twice by the Genetic Engineering Approval Committee (GEAC), the highest authority on the issue.

In India, the level of public awareness of the realizable benefits and probable risks of GE products is abysmally low. The functioning of the GEAC leaves much to be desired. Taking advantage of this hazy situation, mixing up economic, social and political issues with science, and even using such vagaries of nature as the severe drought, several groups of vested interest have created mistrust, confusion and scare. Trashing a very promising technology this way does not augur well for the future of Indian agriculture. This results only in denying the benefits of technology to the farmers and consumers.

The next 20 years of plant biotechnology is expected to improve quality of life, through improved foods, pharmaceuticals, new industrial materials and a better environment. These benefits should reach the people of the developing world, who need them most. For India to benefit from these developments, it is urgent and essential that the public is educated on the realizable benefits and probable risks related to GE products. We also need to reorganize the mechanism of regulating GE products by providing for an expedient, transparent and responsible authority. The media have a very important role to play in this regard. Only an informed and reassured public can make viable choices. Every effort should be made to provide such an opportunity to the consumer.

2

Biotechnology Developments in Agricultural Sector

The Department of Biotechnology (DBT) established under the Ministry of Science and Technology in 1986 was the major instrument of action to bring together most talents, material resources, and budgetary provisions. It began sponsoring research in molecular biology, agricultural and medical sciences, plant and animal tissue culture, biofertilizers and biopesticides, environment, human genetics, microbial technology, and bioprocess engineering, etc. The establishment of a number of world class bioscience research institutes and provision of large research grants to some existing universities helped in developing specialized centres of biotechnology.

There exists a gradient of biotechnologies, depending on the degree of sophistication, complexity, stage of development and application. The lower side of the gradient comprises simpler but widely used techniques such as in vitro culture, rhizobium technology, fermentation. The upper side of the gradient includes advanced techniques involving genetic engineering. The gradient of biotechnologies may be matched with the gradient of national capabilities, economic investment and efforts to provide the possibility of choosing the appropriate techniques and approaches with the most positive impacts. The pattern of use and development of biotechnologies in developed and developing countries thus varies considerably.

THE PATTERN IN DEVELOPED COUNTRIES

In the industrialized countries, a new pattern of biotechnological research funding has emerged. With the availability of property right protection of biotechnologies and prospects of vast markets for biotechnology products and the techniques, the bulk of the research is funded, carried out, and controlled by the private sector. For instance, since 1976 when GENENTECH, the first biotechnology company, was established, the new techniques have spawned a variety of industries that now comprise more than 400 start-up firms, more than 200 established firms that have diversified into biotechnology, and more than 200 supply firms in the United States alone. The new biotechnology industry in the United States produced pharmaceutical, diagnostic tests and agricultural products worth close to US$2 billion in 1990. A similar trend is seen in Europe and Japan. It is estimated that about 60 percent of the biotechnology research and development funding in industrialized countries is from the private sector. Thus, the private sector has been a major force in enhancing the capability of these countries in this field.

Research institutions in the public sector are now generally required to raise a substantial part of their budget from non-government sources, via contractual research, licensing agreements and royalties. This is tending to increase secrecy over research findings and to hinder free scientific communication. University professors, researchers and government institution scientists are increasingly becoming entrepreneurial and are entering private industry.

Another important trend is that large multinational corporations are purchasing smaller seed and biotechnology companies and diversifying their holdings. This allows them to develop a package sale of chemicals, seeds and equipment. Heavy involvement of the private sector and market considerations greatly influence the topics and commodities chosen for research. Major crops, commodities and farming systems of great socio-economic importance to the developing world, but of little international market importance, do not figure in the biotechnology research agenda of industrialized countries. Furthermore, these countries are keen to reduce their production costs, increase the productivity, quality and value of their products and, thus, improve their overall competitiveness in the world market.

THE PATTERNS IN DEVELOPING COUNTRIES

Biotechnology facilities are being established in most developing countries. However, the level of research, development and use of biotechnology for agriculture, forestry and fisheries in the developing countries is generally far below the level in the industrialized countries. Among developing countries, the status varies considerably. A few, such as Brazil, Mexico, India, China and The Republic of Korea, have sought to gain full scientific and technological capacity, especially in agricultural biotechnology. Others, such as Indonesia, Malaysia, the Philippines and Thailand and a few countries of Latin America, have built the capacity to apply biotechniques and develop biotechnologies useful for their agriculture and food industries. The participation of the private sector in gaining biotechnological capacity is not significant in most of these countries.

Many developing countries have inadequate funding, poor human resources and limited access to information, resulting in a relatively low-level capacity for research and technology development and exploitation, especially in the field of modern biotechnology research, which is costly and requires highly trained personnel. Most developing countries are vague as to their immediate aims in agricultural biotechnology. Few have appropriate proprietary-right protection systems or mechanisms to increase their access to protected techniques and products. Furthermore, there is negligible involvement of the private sector, which accentuates the problem of insufficient attention to biotechnology.

One of the major constraints on biotechnological development in developing countries is the poor quality and extent of higher education in frontier sciences, especially molecular biology. Besides, there are no or very weak links between universities and research institutions to reinforce each other and to utilize synergistically the scarce trained personnel resources. The university experts play a negligible role in national policy formulation, including that on biotechnology. Furthermore the universities are generally not oriented for efficiency in commercialization and are therefore not able to market biotechnology products of their own or others' inventions.

Some international and donor-country public sector institutions strive to ensure that attention is given to solving, through biotechnology, the food and agriculture problems of developing countries. Among this group are the International Agricultural Research Centres (IARCs) located or operating in developing countries. Capacity exists in several of these centres for utilization of advances in biotechnologies to solve certain problems of plant production and protection and of livestock production and health, and capacity is being built in others. For example, the International Rice Research Institute (IRRI) is incorporating modern molecular biology in its rice research activities. Its programme is linked closely to the Rockefeller Foundation Biotechnology Network. The World Bank, the International Service for National Agricultural Research (ISNAR) and the Australian Government undertook a major study on the likely impact of modern biotechnology on agriculture. This study is being followed up by the participants by an expansion in the World Bank lending for biotechnology and by ISNAR's establishing an Intermediary Biotechnology Service to provide advisory services to national agricultural research systems.

APPLICATIONS, IMPACT AND POTENTIAL

Biotechnological research and development are moving at a very fast rate. For instance, three years ago, transgenics in rice, especially in japonica rices, were considered rather difficult because of the problem of regeneration through protoplast culture. But today, a large number of transgenics both in indica and japonica types are available and are being tested at various stages. It is therefore difficult to predict the potential and impact of a given technology beyond five years or so. The current level of technology and refers to the developments likely to occur in the next three to five years.

The potential of genetic engineering in medicine has received much attention. The potential for advances in agriculture, forestry and fishery are similarly bright. Biotechniques are already being used to create new strains of crop plants, new plant and animal diagnostic products, animal vaccines, biological pesticides and herbicides, other biological control agents, and modifications in domestic animals used for food production. In some cases, applications are being held in check by the need for still more research to ensure that there are no harmful effects (or, if there are, they are greatly outweighed by the benefits) and by the slow pace of evolution of regulations governing the release of genetically modified organisms or products for general use or acquisition.

Biotechnological research, especially genetic engineering, on problems of field and tree crops, forest species and fisheries has a relatively short history. The knowledge of biological aspects of species and ecosystems needs to be expanded in may cases before new biotechnologies can be applied effectively. Even where transfer of one or more genes can be achieved reliably, expression of the transferred gene(s) may not occur in the expected way. The search for and identification and cloning of useful genes, effective vector systems, methods of gene transfer and promoter mechanisms should be intensified.

A perspective on both the difficulties and intensity of the research effort suggests that diagnostic tools will speed solutions in the next five to ten years to certain disease problems in several crop, forest, animal and fish species, such as black sigatoka of banana and plantain, and virus/viroid diseases of coconut and rice. Improved tolerance to certain physical stresses, such as heat tolerance in potatoes, may also be achieved, but prospects for improved resistance of varieties and clones to most abiotic stresses are longer term.

CROPS

Modern biotechnologies can add greater precision and speed to plant breeding. Transgenics have already been reported in 40 crop plants, including maize, rice, soybean, cotton, rapeseed, potato, sugar beet, tomato, potato and alfalfa, but the new varieties are yet to be used commercially. Near-future opportunities for commercial exploitation include vegetables and fruits (potato, tomato, cucumber, cantaloupe and squash), followed by legumes (alfalfa) and oilseed crops (oilseed rape). A good number of the transgenics are herbicide-resistant plants whose widespread use is somewhat controversial.

Wide use is currently made of tissue culture techniques for micropropagation of elite clones and for freeing planting materials of pathogens.

Monoclonal antibodies are also in use as diagnostic aids in the detection and identification of viruses and viroids. Another culture and microspore culture giving rise to haploids are being used in variety improvement to facilitate and accelerate breeding. Molecular maps and markers are being widely used to identify genes of interest to accelerate conventional breeding programmes. Efficient biological nitrogen fixation systems and strains for efficient utilization of soil nutrients are being genetically engineered. Other long-term objectives are the genetic manipulations of photosynthesis pattern and the production of hybrid seed through apomixis. A very distant possibility is providing nitrogen fixation capacity to cereals.

LIVESTOCK

Among agricultural and allied fields, animal production and health have benefited the most from biotechnology, although practical use of transgenic livestock is only a future possibility. Wide use of monoclonal antibodies for efficient diagnostics, leading to safe and specific treatments of animal diseases, is a major breakthrough. Through genetic engineering, vaccines for the prevention of viral, bacterial, and parasitic animal diseases are rendered more effective and safe. Tailored vaccines exist for pig scours, chicken bursar disease and cattle tickborne diseases. Pathogen-specific vaccines are attractive goals. Other interesting possibilities are endocrine-directed vaccines to stimulate twinning in beef cattle, immunocastration, livestock growth rate increase and vaccines that compensate for various stress-induced production losses.

Advances in genetic engineering will also facilitate the production of male-only populations of screwworms, tsetse flies, ticks and various other ectoparasites for use in the sterile release technique of control or eradication. Furthermore mammalian tissue culture may replace whole animals in the 1990s for toxicity testing of certain chemicals. The culture technique can also be exploited for studying and analyzing pesticide metabolism and for herbicide pre-screening. In vitro fertilization and embryo sexing techniques have considerably increased the use of embryo transfer techniques for cattle breeding and trade. The value of the approach will be further enhanced if embryo cloning techniques can be reliably employed. Microbial and enzymatic treatment of roughages and genetic engineering of rumen bacteria both have great potential to improve animal nutrition. Growth hormones can be produced by genetically engineered microorganisms in quantities and at the low cost necessary for widespread use to speed and increase milk and lean meat production. Biotechnological tools (embryo culture, gene cloning, etc.) may also be used for conservation of genetic resources.

FORESTRY

Successful regeneration through micropropagation or somatic embryogenesis has been achieved for about 100 forest species although, for most, considerable development work would be required before commercial propagation could be contemplated. In breeding programmes incorporating clonal testing, inclusion of a micromultiplication phase may facilitate more rapid deployment of superior genotypes than that afforded by sequential multiplication by cuttings. This will particularly be the case when gene mappling techniques are sufficiently advanced to permit accurate identification of superior genotypes without replicated clonal testing in the field, especially within a breeding line where substantial genetic disequilibrium exists. Furthermore, gene mapping may allow eventual identification of valuable genes, including those that contribute to quantitative traits. As regards the application of genetic engineering, many major traits of commercial importance to forestry are under polygenic control, and much remains to be learned about the operation of the genes involved before this technique can have a major impact for these characters. An earlier application of genetic engineering in some forestry operations may be the introduction of genes, some already available, known to confer insect of disease resistance. Other applications of biotechnology with obvious value for forestry, but as yet not widely supported by experimental successes, include: in vitro manipulation of the maturation state, *e.g.* the promotion of early flowering to reduce generation intervals, in vitro selection for traits such as disease resistance and tolerance to salinity; and the use of haploid cultures.

FISHERIES

Two broad areas of biotechnology exist within the fisheries sector, namely natural products biotechnology (including mostly marine species); and aquaculture biotechnology. Commercially valuable products such as pharmaceutical compounds, pigments, oils, sterols, alginates and agarose are being extracted from micro-and macro-algae in many parts of the world. Aquatic invertebrates are currently being screened for biologically active compounds that may have, inter alia anti-carcinogenic or antiviral properties, whereas marine bacteria are currently utilized in treatments of oil spills and holding tanks in tanker ships.

Within aquaculture, induced increases in the chromosome complement (polyploidy) of commercially important species such as salmon and oysters have increased the growth and marketability of these species. Genetically engineered microbes can be used to produce fish growth hormones, which might then be used to improve feed conversion and growth rate. Synthetic reproductive hormones are produced commercially and used to regulate fecundity, breeding cycles, growth rates and sex determination in certain cultured species. In attempts to increase desirable culture qualities such as growth rate, disease resistance, temperature tolerance and marketability, transgenic fishes containing introduced genes from other species have been produced on an experimental scale. The application of advanced biotechnology to the fisheries sector is a relatively recent practice. Therefore, continued research can be expected to reveal additional commercially viable applications and products within this sector.

BIOTECHNOLOGY IN AGRICULTURAL DEVELOPMENT

We are at the threshold of 21st century, where great challenges on three fronts namely, production, pollution and population are awaiting the developing countries. Moreover, in the era of privatization, globalization and liberalization, there is heavy international competition within as well as between the countries on all fronts including the agriculture. The present scenario indicates that there would be "survival of the fittest". Needless to mention that by 2000 AD our population will be around one billion the priority is clear-cut before us. We have to climb upward to achieve higher and higher production to ensure that every mouth of our ever increasing population has enough to eat.

The quest for agriculturists for greater productivity and improvement of existing cultivars with better quality food continues with great vigour and spirit. The conventional methods of plant breeding and traditional agricultural practices have done a tremendous job and contributed to a great deal towards the above goal. However, in view of the acuteness of the problem and renewed fears regarding the availability of the proper and enough food, these methods alone are not sufficient to meet the situation.

Although, the FAO believes that food production will continue to increase for the next 20 to 25 years at the same rate as experienced during the past 30 years to meet the futuristic demand. The worldwatch Institute has, however, reported that this rate may not be sustainable. Moreover, they believe that science and technology can no longer ensure the onward march of achieving higher and higher yields. In addition we are also subjected to an almost daily litany of doom and despair as a result of global warming, depletion of ozone layer, oil and water running out and so on. The next century certainly packed with trouble if the meteor does not get us first, then we will gradually either be roasted, frozen or desiccated, and to make doubly sure, shall certainly be starved as well.

It seems that if we have to stand any chance whatsoever in the next century, our capacity to adopt the living resources will be extremely important for our survival. Because of changes in the environment and depleting energy resources, this capacity to adapt will be every bit as important to those areas of the world which are blessed with overproduction like Europe today. Accelerated adaptation through genetic change is, of course, one of the important means by which plant breeders have achieved their aims in the past. Today the new techniques of biotechnology in general and genetic engineering in particular may offer to the plant breeders and agriculturists the chance of speeding up adoption to an extent hitherto considered impossible. This new power will undoubtedly bring risks but it could also be the only chance we have for escaping the dangers ahead. Biotechnology being a multidisciplinary subject requires coordinated efforts and expertise of scientists with different backgrounds to put together in order to achieve success. It may be defined as may technique that uses living organisms, or parts of organisms to make or modify products, to improve plants or animals or to develop microorganism for specific uses.

It includes well established technologies such as those used in breeding, plant propagation and conventional animal vaccine productions. Modern biotechnology encompasses the more recently developed techniques involving the use of recombinant DNA technology, monoclonal antibodies and new techniques of cell culture. Recent development in biotechnology have made it possible to move genes from microbes, plants and animal species into plants of interest. Biotechnology helps in the commercial utilization of gene and other things for the benefit of human kind. However, destruction of ecosystems has been a major concern of biotechnology.

Developing countries are naturally attracted to the potential applications of biotechnological research in solving problems of hunger, energy supply and improving the quality of life. The priorities of different countries however very widely, In this context, the National Biotechnology Board of India has chosen genetic engineering, photosynthesis, tissue culture, enzyme engineering, alcohol fermentation and immuno-technology as areas of immediate interest.

APPLICATION OF DNA TECHNOLOGY

Genetic engineering opens a totally new dimension for bioprospecting. The search for new genes and their application is the primary objective of the biotech industry. Gene technology now enable humans to integrate revolutionary new properties in to cultivated plants through inter-specific or inter-generic gene transfer which was not possible through classical approach of crop improvement.

Genetic engineering is the direct introduction into a plant of an isolated or modified single gene using transformation techniques. Now it is possible to transform almost all of the plants cultivated by man. Where this has not happened, it is more to do with the insignificance of the plant in agriculture or forestry rather than a reflection of a stubborn resistance to all attempts to transform it. A couple of years ago, cereals were regarded as being recalcitrant species. This has new changed and transformation has been reported as being successful for nearly all cereal species.

TRANSGENIC PLANTS

Transgenic is used to describe plants which have had DNA introduced into them by means other than by the transfer of DNA from a sperm cell to an egg. Development achieved through modern biotechnology in the form of transgenics have given new options of controlling insects, diseases and weed pests, while improving the overall integrity and consequences of agricultural practices. Biotechnology combined with modern crop management techniques and the responsible use of pesticides gives the best tools available to ensure healthy, high yielding crops. However, the benefit expected from the release of genetically modified organisms into environment are quite more which are seems to pay major role in bioremediation, environmental improvement, agriculture, food industry and health care. A number of transgenic plants in various crop species covering wide range of altered characteristics have been approved for commercial cultivation.

RESISTANT TO BIOTIC STRESSES

Damage to crops by biotic factors like insects-pests and weeds is a major limiting factor in agriculture economic in tropical and temperate regions of the world. Despite large scale investment in the chemical control of pests, they cause great economic loss by damaging or destroying the crops. With a tremendous development in techniques to engineer crops genetically, we now have the ability to make broader use of natural insecticides. The choice of a Bt endotoxin, as the first insecticidal protein for introduction into plants, was based on the extensive knowledge gathered about this class of crystal protein since 1902 and such a strategy has been successfully achieved. The application of glyphosate kills crop plants just as effectively as it kills weeds. The usefulness of glyphosate as a weed control agent in agriculture could be enhanced if resistance to herbicide could be selectively engineered into crop plants. Transgenics have been developed by introducing the corresponding resistance genes from weeds and microbes into crop plants such as tobacco. In addition genes conferring resistance against herbicides, *viz.*, Sulphobylurea, Imidazolinones, Phosphinothricin, Asulam, Bromoxynil and 2,4-D have been engineered and transgenic plants developed.

ABIOTIC RESISTANCE

A large area of our country is under stress conditions *i.e.* saling, alkaline water, cold and heat stress etc. transgenics produced from introducing mtl 1D gene from E. coli in tobacco and arabidopsis showed tolerance to high leave of NaCl. Thermoc tolerance was found in Arabidopsis plants engineered with hsf gene. It has been reported that over-expression of sac B gene from Bacillus subtillis leads to high level of fructans in tobacco cells, and this is associated with increased drought tolerance whereas chemically synthesized antifreeze protein genelala 3 leads to improved freezing tolerance of tobacco plants.

Genetic engineering for improved tolerance to abiotic stresses is, therefore, the need of the hour as the existing cultivars in most cases are capable of giving much higher biomass that what we harvest today. While it is true that the response of plants to these stresses is multigenic, the recent success achieved in genetic engineering against excess salt, high and low temperature as well as less or excess water by altering individual gene is noteworthy.

PRODUCT QUALITY

A group of scientists have reported the ethylene synthesis in transgenic tomato plants, using anti-sense to a gene that has been identified as ACC oxidase. The enzyme convert ACC to ethylene. Ethylene production was inhibited by 97% with a signification reduction in over-ripening and shrivelling. Flavr Savr and Endless Summer in tomato, Freedom II in squash, high lauric acid reposed (Canola) and Round Up Ready soybean are some examples of the crops that already being commercially grown in developed countries.

Genetic engineering of metabolic pathways of fruits and vegetables has therefore given control over post-harvest process which not only leads to improvement of quality characteristics but also enhances processability, transportability and prologation of shelf like.

In addition, engineered plants with change fatty acid composition of edible oil, reduced antinutritional components in many food crops, production of palm oil or animal derived industrial oil into plants like Brassica are some of the good examples of novel products that can be expected from biotechnology. Significant progress has been achieved for production of biodegradable plastics derived from biomsass of genetically altered plants having gene from bacteria. With the advancement in DNA technology and understanding the structure and function of gene and its products more and more transgenics are coming out from various research programme all over the world. However, boon of biotechnology has been confined to few developed countries and developing countries are still in vogue to harvest the benefits of modern technology. More that 48 transgenic has picked up remarkably in last couple of years. During 1986-87, 25000 field trials were conducted with transgenics of more that 60 crops in 45 countries. Sixty per cent of these trials were in first 10 years and 40% in the last two years. Area under transgenics cultivation has also increased tremendously. During 1996, the transgenic crops covered an area of 2.8 million has which increased about 4 times during 1997 and 10 times during 1998.

PRODUCTION OF TRANSGENIC PLANTS

To produce transgenic plants, the crucial steps are the introduction of the cloned genetic material into the plant cell nuclear & the facilitated integration of the cloned gene into the plant chromosome. It is interesting that the best method for doing so uses a system that already exists in nature, the system by which the plant pathogenic bacterium Agro-bacterium tumefaciens injects a portion of its plasmid DNA into plants & inserts it into the plant become. The Agro-bacterium system is an excellent system for introducing foreign genes into the chromosomes of plants. The transfer into intact plant cells occur at high frequency, the T-DNA is usually integrated into the plant chromosomes at high frequencies without undergoing structural alterations, and the cells that have received the T-DNA can be selected easily by using antibiotic resistance such as the neomycin resistance maricer. Finally the transgenic plants produced in this manner are quiet stable for at least several generations. Consequently almost all existing transgenic plants with potentially desirable traits have been obtained by this method.

HERBICIDE RESISTANCE PLANTS

The advantages of making crop plants resistant to herbicides are obvious. Although there is a fear that such resistant may eventually increase the use of herbiude chemicals, there are also reasons to except that these transgenic plants will promote the use of sager, more biodegradable herbicides, perhaps in smaller amounts one example involves the herbicide glyphosate which inhibits 5-enol-pyruvylshikimate 3-phophate synthase-an enzyme involved in the biosynthesis of aromatic amino acids.

This enzyme has been purified from crop plants & sequenced and DNA Probes corresponding to its amino acid sequence have been synthesized. These probes were used to isolate CDNA for the enzyme from the CDNA library of a plant cell line known to overproduce 5-enol-pyruvylshikimate 3-phophate synthase. The CDNA was then cloned behind the strong CaMv promoter, and the promoter gene complex was introduced into plant cells via a disarmed Ti plasmid vector. The transgenic plants produce a much higher level of the target enzyme & thus are significantly more resistant to glyphosate. These results are encouraging because glyphosate has very low toxicity to animals is rapidly degraded in soil.

INSECT RESISTANT PLANTS

Bacitlus thuringiensis is used in the biological control of caterpillars because its sporulating cells contain toxic proteins. The gene for one toxic protein was cloned behind promoters that are effectively expressed in plants & was introduced into plant cells via a Ti plasmid vector, thus producing plants toxic to caterpillar. The major problem with this approach has been the low level of expression of the toxin protein in plants, which is presumably related to the fact that the gene come from a bacterium,. Nevertheless, the method has produced tomato & tobacco plants that proved quiet resistant to caterpillars in field tests. Cotton however, is often attacked by insects that are more resistant to the B. thuringiensis toxin and the low level of expression of this toxin in transgenic cotton plants provided the plants with little if any protection. Recently, the coding sequence of the toxin was altered extensively to replace codons that are rarely used in plants as well as to preclude the formation of a strong secondary structure in MRNA. When cotton plants were provided with this modified gene, their production of the bacterial toxin increased 100-fold and they showed impressive resistant to common bepidopteran insects than damage unmodified plants.

Some seeds contain high concentrations of protease inhibitors, which are thought to interfere with the digestive process in insects. The cloning of cowpea trypsin inhibitor genes, for example and their transfer to tobacco plants have resulted in good resistance to a wide variety of leaf-eating insects.

VIRAL RESISTANT PLANTS

The method most frequency used for producing virus-resistant plants sprang from the observation that, of ten times, plants infected by nearly avirulent virus are thereafter resistant to super infection by a related, highly virulent one. Thus, deliberate infection with avirulent virus strain has been used to produce protection in crop plants. The method is not totally safe, however, because the avirulent strains many mutate to produce strains that are significantly pathogenic. Most plant viruses are positive strand RNA viruses covered by coat protein sub units. When the virus enters an injured plant cell the replication process begins, starting with the progressive uncoating of the virus from the 5-end of the RNA. Tranagenic plants, whose genomes contain introduced tobacco mosalc virus (TMV) coat protein genes and which continuosly synthesize the coat proteins, show resistance to virus infection. The phenotype of these plants is consistent with the idea that resistance is a result of interference with the uncoating of the virus particle. Although the plant cells are resistant to infect TMV particles, they remain sensitive to the TMV RNA or to partially uncoated TMV particles.

The coat protein gene is usually put behind a strong promoter such as the 35S promoter of CaMV & is introduced into a plant's genome via the Agro-bacterium Ti system. Transgenic plants showing significant resistance to TMV alfalfa mosaic virus, cucumber mosaic virus, tobacco streak virus & tobacco rattle virus have already been produced in this manner. In a recent experiment two virus & potato virus Y, were introduced simultaneously into a commercially important potato cultivar, and one of the resulting trangenic plants provide quiet resistant to both viruses under field test conditions.

BIOTECHNOLOGY FOR THE COTTON FARMER PROBLEMS AND PROSPECTS

Cotton or white gold as it is aptly called is grown for its lint and seed which yield cotton fibre and seed oil, respectively. This crop occupies 32-33 mha of world area with a production of 20-25 metric tonnes. In India its area spans over 7.5 mha with an average yield of 290 kg/ha of lint and 870 kg/ha of seed cottom. To meet the challenges of 2000AD with a population of more that 960 million, a total production of 19 million bales is required as against the 13-14 million bales of today. This can be achieved by the use of improved crop production practices coupled with appropriate pest management tactics. In addition, generation of novel. Transgenics may help achieve the near impossible.

Genes that have been identified as potentially profitable, if engineered into acceptable cultivars can be used to generate such transgenics. Among these are genes imparting resistance to herbicides, insects, pathogens and abiotic stresses. It is also widely accepted now that a number of other qualitative characters can be improved, such as fibre strength, fineness, colour and thermal adaptability of the fibre.

Thansgenic plants have become realistic components of stress management world over.

Bollworm and herbicide resistant transgenic cotton have received the approval of the Environmental Protection Agency and have been commercially released in the US for cultivation.

Considering the fact that numerous biotic and abiotic stresses limit cotton production, it is likely that future strategies might orient towards the development of a multiadversity resistant high yielding transgenic cotton variety with superior fibre qualities.

The most important aspects in the development of transgenic plants are

* Identification, isolation and clining of the desirable genes
* Transformation of plant cells or tissues with a suitable vector
* Regeneration of the transformed cells or tissue
* Confirmation of the gene integration and expression
* Hardening and field establishment of the transgenics.

Major achievements worldover in the field of cotton transgenics

GENES

Genes for Insect Resistance Used

* Toxin genes from Bacillus thuringiensis
* Protease inhibitor genes from plants & the horn worm

Genes used for herbicide resistance :

* Nitrilase gene from a bacteria Klebsiella spp. for Bromoxynil resistance
* Mono-oxygenase gene from a bacteria Alcaligenes eutrophus for 2,4-D resistance
* Acstolactate synthase gene for resistance against sufonyl urea & imidazoline group
* ERPSP synthase gene for glyphosate resistance.

Genes used to Improve fibre qualities :

* Polyhydroxybutyrate & polyhydroxyacetate from bacteria
* Expression of indigo pigment in fibres

Genes to be Used

* Cholesterol oxidase genes against insects
* Heliothis stunt virus genes against Heliothis
* Genes encoding chitinases glucanases, attacins cecropins vital coat proteins etc. against diseases.

Transformation

* Agrobacterium medicated transformation
* Particle acceleration gene delivery systems have been successfully used all over the world.

Regeneration

* Callus fissue regeneration-mostly genotype specific
* Protoplasts were regenerated successfully.

Meristematic tissues were successfully regenerated.

Expression

* Enhanced expression was achieved using truncated forms of the full length genes, repeat copies of the propoter, modification of the bacterial codons to plant prefered type.

Commercial Release :

* Bollworm resistant transgenics-NUCOTN 33M & NUCOTN 35B were released in the US and cotton transgenics-are in now under field cultivation in Australia, South Africa and China.
* 8 lakh hectares were planted with the transgenic cottons in 1996 and 1997
* More than 8,000 hectares were damaged by bollworms.
* Transgenic cotton plants in Australia were also found to be attacked by the bollworms.
* Herbicide (Bromoxynil) resistant transgenics BXN 57 and BXN 53 were released in the US in 1996.

Problems Encountered/Expected

* Gene silencing and expression instability
* Resistance in insects to the toxins used. Studies on baseline toxicity indicated that there is a natural variation in the ability of Helicoverpa armigera in tolerating the CrylAc toxins. Some strains are capable of surviving the totoxins by virtue of an in built capacity to tolerate the toxins, while others are not. While development of Helicoverpa resistance to the CrylAc toxins itself may not be of immediate concern, a slight shift in the tolerance levels resulting from continuous exposure to the transgenics, can cause control difficultes.

* High cost of transgenic seed.
* Microecological changes resulting in pest shifts. CrylAc is a broad spectrum lepidopteran toxins that can cause significant changes in almost all populations of lepidopteran insects occurring on the cotton crop. Some of these insects, which cause negligible economic damage to the crop, harbour parasitoids which have the potential to keep the bollworm populations under check. Consequently, in the absence or low populaions of natural enemies, the bollworms can emerge as stronger pests that before. Insects pests such as Spodoptera litura which are less affected by the CrylAc toxins and which have been under check due to the use of pyrethroids can resurgace as major pests as pyrethroid use is likely to be reduced on transgenic Bt cottons.

Some Recent Development in Cotton Transgenics

* Three Bt transgenic lines, 95-1, RIOI and R-19 have been indigenously developed in China through Agrobacterium medicated (Somatic embryo) and pollen tube pathway transformation. Thestability of Bt gene expression and field efficacy have been confirmed.
* Microprojectile bombardment parameters for pollen mediated transformation and genetype independent protocols have been develop in China.
* A new methods of transformation through injection of Agrobacterium into developing embryos, has been reported from South Africa.
* Stable transformation and regermation has been reported for Bt transgenics in Uzbekistan, China, Egypt and Australia. Pakistan reports development of cotton transgenic plants resistant torearcurl virus.
* An antisense DNA of CLCUV DNA-A bome ACI gone along with the antesense DNA of the AC2 and AC3 gene was used for the vector construction and transgenic cotton resistant to the CLCUV was developed in Pakistan.
* In Russia, A glucanase gene has been isolated from thermo philic bacteria and is being used as a new reporter system fortransgenic cotton development.
* Genetically engineered Baculovirus ACMNPV with genes from straw itch mite has been developed been developed by Monsanto and Zeneca with researchers from Madison, USA.
* Bt rransgenic cotton currently occupies 20% (85,000 ha) of cotton area in Australia. Isecticide use has been reduced by about 60% The efficacy has been moderate. The transgenic plants were found

to express Bt proteimns only till the 95th day after which fluctuaions were observed. The reduction in expression was primarily due to down regulation which post transcriptional changes of the unstable RNA It was also reported that cotton tannings which increase with growth phase, act as antagonistswith Bt toxins.

* After two years of field cultivation, true resistance to Bt has not yet been detected in Australia.
* Reslistanor to Bt in field populations of Helicoverpa armigera is reported form China
* Annually 100 million dollars are now being spent in the US, only in search of new insecticidal genes.
* In the US, 45% of cotton area this year will be under transgenic cotton (herbicide and insect resistant). However, the area under Bt cotton seems to on the decline.
* Transgenic cotton developed in Australia using the Stunt virus genes, is not performing well.
* Transgenic cotton developed in the US using lectin genes, is not performing well against Heliothis virscens.
* The Australian report that Helicoverpa armigera is at least 100 times less susceptible to Cry toxins as cinoared to Heliothis virescens.
* The Window strategy for insecticide resistance management in Australia is still important for transgenic cotton.
* Cotton Pest spectrum in the US is getting altered after the introduction of Bollgard. Unsprayed Bt cotton sustained 4 times more attack of tarnished bugs, 2.4 times more with boll weevil, 2.8 times more with stink bugs and Spodoptera.
* Due to these changes in pest complex farmers had to spray 3-5 times on boligard as compared to 6-8 times on non-Bt cottons.

PROSPECTS AND POTENTIAL

The application of biotechnology in cotton farming can be either in the form of production of fermentation products or novel recombinant products for use (example, recombinant Pseudomonas expressing CrylAc) or as transgenic plant with in buitt resistance to biotic and abiotic stresses. Transgenic crops with in built resistance to insect pests and diseases can be extremely useful as this would result in the reduction of insecticide use apart from making pest management simple for the farmer. Introduction of the bollworm resistant transgenic cotton is expected to reduce the use of chemicals used to control bollworm especially Helicoverpa importantly at a time when resistance to most insecticides including pyrethrolids is on the increase all over the world. Even if the introduction of the Bt cotton in India could result in a 25-30% reduction in insecticide use on cotton, this would mean a benefit of about of about Rs 300 crores, apart from the f-avourable impact on the environment so far, transgenic plants have been produced in about 60 plant species. Cotton has received special attention of the biotechnological companies in the developed countries who were attracted by the profit motives associated with the high value added to the transgenic seeds.

The firs Bt transgenic cotton has already been released, as Bollgard. The Delta and Pineland using DP 5410 and DP 5690 as recurrent parents. The D and PL brand BT transgenic were labelled as Nucotn 33B Two million acres were planted in 1996 in mid-south region of USA, mainly in regions where Heliothis virescens was problematic. Nearly 1.2 M acres were under Nucotn 33 B and the rest under Nucotn 35B.

The Bollgard brand Bt cotton seed was sold at US $34-36 as compared to US $8-9 per hectare for non-engineered cotton seed. The average cost of control of cotton insect pests in the US was approximately US $150 in the early 1990s. Hence the prices were still found to be attractive. The transgenics were found to be more effective against Helothis virescens as compared to pectinophora gossypiella and Heliothis zea.

Farmers growing transgenics had to sign'an agreement with D & PL stating that they would not keep seed for planting next year. Quick ELISA tests have been devised to test for Bt toxins in plant parts, to check the illegal spread of transgenics.

Two more transgenic cotton varieties tolerant to herbicide Bromoxynil, trademarked as BXN 57 and BXN 53 were developed from the recurrent paren Coker 315 BYTHE Stonevuille pedigree company in collaboration with Calgene Inc. This seed was priced nearly 1.7 times the straight varieties.

In Australia the CSIRO and Narrabri research centre, in collaboration with Monsanto have developed transgenic cotton plarts that were commercial released in 1997. This introduction resulted in 50-60 % reduction in the US $ 93 million spent by farmers each year on insecticides. Pakistan which uses 90% of its total insecticides on cotton alone and China which is one of the major consumers of insecticides on cotton are reportedly strongly considering the prospects of introduction of Bt transgenic cottons.

Cotton acreage in India is only 5% of the total cropped area, yet it consumes more than 50% of the total insecticides used in the country. The pattern of usage, however, is not uniformly, spread, for instance in Andhra Pradesh alone, where cotton cultivation is only 0.3% of the total cropped area of the country, accounts for 17% of the pesticide use on cotton in India. Again within the average picture of the state, coastal AP uses 30% more pesticide than the state average. The approximate estimates of insecticide sales in Guntur district alone, where cotton is grown in 0.15 mHa, is about US $ M. Clearly areas with maximum pecticides use per hectare, necessitated mostly due to bollworms, are likely targeted market niches for the bollworm resistant transgenic cottons. Moreover, since the price of the transgneic seed in the US is nearly four times the high input cotton cultivators of the irrigated belt Growing Bt transgenics and herbicide tolerant transgenics on marginal land is not recommended by the company as these would not show any impact in the absence of strong insect pressure and broad leaved weeds. In most areas of rainfed cottons the use of insecticides and pest pressures are low hence, it may not be economical in such belts, to encourage the use of transgenics.

Now transgenic production technology for many crops has become an established routine procedure in many countries including India. The research on transgenic cotton in government funded research labs in India is almost in the final stages. Meawhuile Mahyco-Monsanto biotech is ready to market Bt transgenic cotton (bollgard) in India. The introduction of transgenics through these companies are to be in the form of FL seeds wherein Bt genes are inherited from the transgenic exotic germplasm. Such seeds may also be priced at a premium and it remains to be seen how enthusiastic the response of Indian farmers would be to the expensive input. Transgenic releases from government organisations would definitely enforce a competitive pricing to restrict the monopoly of the private companies.

The transgenic technology is also likely to be extremely beneficial to the private sector. For instance, a transgenic with resistance to a specific herbicide marginalises the use of other herbicides for chemical weed control and the chemical company holding the patent on the herbicide might be prove to be beneficial to the resource rich farmers. Clearly, here the targeted market is primarily high input farmers. The monopoly of transgenics due to patenting rights held by the biotech companies may also lead to high pricing of the seeds.

A Significant socioeconomic issue that arises from the introduction of transgenics into the Indian farming system is that the high priced seeds may benefits the prosperous and large farmers thus providing a negative externality on small and marginal farmers. On the other had it is also argued that the developments from the application of biotechnology would be beneficial to low input farming practises wherein the cost of chemical inputs can be minimised. It is now only a matter of time before we experience the full social. economic and environmental impact of transgenics in our country.

SCIENTIFIC AND INDUSTRIAL RESOURCES FOR BIOTECHNOLOGY

As far as scientific and technological production in Latin America, university–industry research relations are almost nonexistent. Universities in Latin America are more oriented to learning and teaching than to doing research. One might say that the only practical relation between universities and industry is that of supplying the latter with a qualified labour force for professional and operative work. For the sake of illustration, let us look briefly at the case of Mexico, where university research is highly concentrated. The

National Autonomous University of Mexico (UNAM) alone concentrates 50% of all research funds in the country. Nevertheless, UNAM's research budget represented only 22% of its global budget for 1988. The rest represented wages and other operating expenses.

The Mexican dilemma is that its science and technology system produces so little technology that most of it has to be imported. Still, there are four categories of university–industry links: training of professionals, creation of basic and applied knowledge, provision of technical services, and production of technology. Part of the explanation for the weakness of university–industry links in Mexico is the near absence of a commercial criterion in research. Thus, the prevalent criterion tends to be defined in epistemological terms, which may be crucial for science but irrelevant for technology. For technology development, what matters is economic convenience, not epistemological relevance.

There are, however, some institutions doing cutting-edge biotechnology research in Mexico. In total, there are about 25 centres with biotechnology projects broadly defined. If we narrow the definition of biotechnology to the newer techniques, such as those based on recombinant DNA, the generation of hybridomas, cell and tissue culture, and protoplast fusion, then the number of institutions is reduced to only seven. The rest are involved in traditional biotechnology research. Of those seven, three belong to the National Autonomous University of Mexico, two to the National Polytechnical Institute, one to the Autonomous Metropolitan University and another one to Chapingo Autonomous University. All of these universities are located around Mexico City, although three of the top research centres involved are decentralized in neighbouring cities (one in Irapuato and two in Cuernavaca).

The pioneering institution in promoting industry links has been the National Autonomous University of Mexico (UNAM), reputed as the largest university in Latin America. It created a "Centre for Technological Innovation" in 1984 to market UNAM's science and technology. Although the explicit objective of the new centre is to market UNAM's science and technology, a former director declared to Expansión, the Mexican version of Fortune magazine, that they wish to preserve their identity as an academic institution. They are not after tailor-made research, therefore, for the specific needs of companies. Rather, companies should find out what UNAM's researchers are doing that might be of interest to them. The director of one of the firms that has supported UNAM's research stressed that business firms should take advantage of Mexican technology, because it is cheaper and should allow them to compete better in international trade.

In sum, Mexico does have several excellent research groups working in the frontiers of biotechnology. But the number of researchers is very small in comparison to those in advanced capitalist countries, both in universities and in firms. In developing countries, there tends to be a very weak if not totally absent link between universities and industry. With such great differences in the quantity of scientists and the differing patterns of university–industry linkages, it is hard to be very optimistic about the possibilities for developing countries to compete successfully in such a dynamic field as biotechnology on the basis of their own resources.

It is necessary, therefore, to undertake a more realistic analysis regarding the actual scientific potential of developing countries and what public policies it would take to translate research into products that might solve their specific socioeconomic and environmental needs. Unless the proper policy incentives exist, it is unlikely that these needs will be taken into account by the private firms doing biotechnology research, or by scientists from advanced countries who have other concerns as their central motivations.

ECONOMIC SCALE AND BIOTECHNOLOGY

Contrary to what optimistic analysts of biotechnology hold, it seems that the main tendencies indicate a bias favouring the large scale, rather than scale neutrality. That is to say, far from favouring small-scale farm operations, the new biotechnologies will tend to reinforce the bias introduced by modern agricultural technologies by requiring strong capital investments. With the new technologies then, the agricultures of developing countries are likely to require a smaller work force, even if we assume that many of the products of biotechnology may be adopted by those countries.

The alternative might be that, if new technologies are not adopted, and given the global scope of today's world market, agriculture in developing countries may cease to be a viable economic activity altogether. This would be a major paradox, for about 60% of the population in developing countries depends on agriculture. The following examples of new biotechnology products clearly reflect an economic bias towards large-scale production units. These are the implications of the "technological paradigm" represented by modern agriculture.

The generation and sale of cattle and sheep embryos is generally reserved for large and modern production units, which utilize the most advanced genetic technologies. Commercial cloning of livestock is becoming a reality. There are research groups in the U.S. that are developing cell fusion methods with the potential to create multiple, perhaps unlimited, copies of individual cow-and sheep-embryos.

The target in cloning is to produce genetically predictable animals, with superior milk-and meat-production capabilities. University of Wisconsin Professor Neal First and colleagues have successfully cloned cow embryos, but it will be years before they can determine the quality of the resulting breed. For what they clone is not the mother cow but the embryo itself, *i.e.*, an already fertilized egg. One of the goals of researchers is to improve the cloning technique to develop the entire process in test tubes and eliminate the need for embryos from the cows.

One implication of this type of research is that, in the future, it will be possible to produce "elite" animals for meat and milk production. This genetic improvement of animals, combined with the use of growth hormones, promises to revolutionize production in these fields, and they will probably displace large numbers of farmers, given the large-scale bias. The dictum for this type of modern operation will continue to be, perhaps in an exacerbated manner, "get big, or get out of farming."

Socially, this trend points to a future agriculture that will require not much more than a percent or two of a country's labour force. The main concern will be with how to employ productively, with adequate incomes, those people who are displaced from agriculture. In the U.S., this stage has already been reached: only 2% of the labour force is directly employed in farming.

Another factor that reinforces the previous interpretation regarding the economic bias towards the larger scale is that new agriculture will probably combine two or more products of the current technological revolution, including computers, for the integrated management of agricultural enterprises. It is very likely that a systems approach to management will be reinforced. For instance, and continuing with the example of livestock production, the most sophisticated dairy operations may use embryo transfer technologies to improve the herd, while using computers to maintain exact yield records per cow and to monitor their feed requirements.

The result will be that only large-scale operations will be able to make the heavy capital investments required for such integrated systems for managing resources. All of this reinforces the economic-scale tendency to discriminate against small-scale producers and accelerates the process of "depeasantization without full proletarianization". The dilemma involved in this process is that, although farming expels many of its workers because of increased capital-intensity, the larger economy cannot absorb all those redundant workers. This is the source of the swelling "informal sector" in Latin American economies.

STUDYING BIOTECHNOLOGY'S FUTURE IMPACTS

Because agricultural biotechnology products are still largely in a laboratory stage, there is room to hope that some social forces will be able to steer their development in more desirable directions. This calls for enlightened policy by governments and international agencies that must be ahead of the competitive dynamics that drive businesses and technological innovation.

A tough call indeed, for governments must function within the confines of parliamentary committees or government agencies, whereas companies and researchers follow a much swifter market temporality. It is between the logics of state and markets where sociological analysis can help identify areas of social organization where some policy and action may be effectively directed. A new approach to study the emerging dynamics of global capitalism and suggests ways in which it could be utilized in the assessment of future impacts of biotechnology. There are actually two approaches that may be combined in studying the future impacts of biotechnology. One covers the global dynamics of capitalism, the "global commodity chains" approach (GCC), and the other might be termed "bottom-up linkages" (BUL) approach. The GCC approach addresses the links among households, enterprises, and states in global production, whereas the BUL approach highlights the importance of local conditions and social actors in future technological innovation.

With these two approaches one can move beyond and below the analysis of the nation state in studying the course of development. Let us begin with a brief description of the GCC approach.

William Friedland (1984) first proposed the "commodity systems" approach to understand the dynamics of global capitalism with regard to agricultural commodities. More recently, the contributors to Gereffi and Korzeniewicz (1994) developed a " global commodity chains" approach that is complementary to that of Friedland.

On the one hand, as Laura Raynolds puts it, Friedlands analysis "focuses on the organization of production (including production scale, labour organization, and the role of science and technology) and its integration into marketing and distribution systems". On the other hand, the commodity chains approach, a term first proposed by Hopkins and Wallerstein (1986):

emphasizes the interlocking "nodes" of production that go into creating a finished commodity. This latter formulation is more sensitive than the former [Friedlands] to the links between component production processes, the geographical location (and potential dispersion) of production, and the variability of "commodity chains" over time. Yet Friedlands approach more carefully avoids the reification of commodity systems and places greater emphasis on microproduction relations. In the case of future biotechnology products, one initial task would be to identify the key global commodity chains that will be affected to concentrate research on those that are regarded as the most relevant by social and environmental criteria.

A global commodity chain (GCC) "consists of sets of interorganizational networks clustered around one commodity or product, linking households, enterprises, and states to one another within the world-economy." Furthermore, this type of analysis, shows how "production, distribution, and consumption are shaped by the social relations (including organizations) that characterize the sequential stages of input acquisition, manufacturing, distribution, marketing, and consumption".

Some key concepts of comparative sociology, such as national development and industrialization, are increasingly perceived as problematic with this multilayered approach. For the locus of industrialization may no longer be an advanced capitalist country but a developing country, yet the control over the GCC continues to reside in the wealthier countries whose service industries seem to thrive.

Viewed in a global context, however, it becomes clear that the service sectors in wealthier countries are firmly linked to other productive sectors in a GCC, which shreds the thesis of postindustrial societies to pieces. In other words, manufacturing continues to be at the root of global capitalist dynamics, and the so-called service industries also entail transformation to the extent that they are based on human skilled judgment.

Other important distinctions that have been elaborated by the GCC approach regard the dominant force within each commodity chain. This refers primarily to the type of market in question: is it an oligopolistic or an oligopsonic market, that is, one dominated by a few producers or by a few buyers? Oligopolistic GCCs are called "producer-driven" chains, whereas those dominated by a few buyers are called "buyer-driven" chains.

This distinction is important in the case of agricultural production, even though most commodity chains are producer driven. In the agricultural inputs sector, for example, the agricultural machinery companies or the petrochemical companies largely determine the technologies supplied for agricultural production. They dominate their backward linkages with their own suppliers and, because the industry is highly concentrated, they also dominate their forward linkages with their consumers, the farmers.

Furthermore, there is the prevalence of contract-farming arrangements by producer firms that also act as a monopsony. For example, Campbells controls about 90% of the canned-soup market in the U.S. and is the single buyer of a number of agricultural commodities, namely tomatoes, from thousands of farmers. Similarly, many agricultural commodities such as many grains tend to be dominated by commercial capital or "buyer-driven" commodity chains. Witness the domination exercised by companies such as Cargill in the soybeans world market.

Whether biotechnology products will be channeled through producer-driven or buyer-driven GCC will likely involve a different combination of policies in trying to shape their social and environmental impact. For instance, Rodolfo Quintero (1992) recently suggested that tomatoes embody such a large relative value and are such an important export crop for Mexico that biotechnology research efforts should focus on this particular commodity.

Although these facts may be true, we should also ask the social question of who controls this commodity chain and to whose benefit would such research efforts accrue. It is not enough to do an economic calculation in terms of value added and trade balances; one also has to study issues of employment, whether small-, medium-, or large-scale farmers dominate within each commodity chain, and the type of environmental practices that prevail.

Questions should also be raised as to the type of GCC that dominates the agricultural inputs sector that supplies a given commodity chain. Will it be possible to establish a new marketing network in a commodity chain with already firmly established agrochemical firms, *e.g.*, Monsanto? Or would it be best to seek an alliance with such companies, trying to reshape their research agendas in more benign directions? With regard to the "bottom-up linkages" approach, the key question is the extent to which local environmental and social conditions are taken into account in technology development. With conventional agriculture, scientific approaches have come to dominate the generation of technology, largely abandoning local knowledge and practices. Just as much biodiversity has been lost to the planting of homogeneous modern plant varieties, so has much local and indigenous knowledge been wiped out by the sweeping introduction of conventional agriculture in developing countries.

In the case of Latin America, our scientists tend to be trained in Northern countries and they adopt the technological paradigm of conventional agriculture. There has been much disregard, therefore, for building on the local conditions and local knowledge of farmers when introducing technological innovations.

With the hegemony of neoliberalism and the dominant trend towards the globalization of the world economy, the forces of conventional agriculture can only get stronger.

But public concern for social and environmental sustainability also runs high. It will take democratic politics and decision-making, therefore, to address these concerns beyond the dictates of the market and profitability concerns. Only if public institutions, including universities, and grassroots farmer organizations establish an alliance can there be any hope to move into a more desirable form of development.

3

Genetic Biotechnology in Agricultural Sector

The genetics and genomics revolution has at its core information and techniques that can be used to change humanness itself as well as the concepts of what it means to be human. The age-old human fantasies of the mythical chimeras of the ancients, supernatural intelligence, wiping disease from human inheritance, designing a better human being, the fountain of youth, and even immortality now have biotechnical credence in the theoretical promises of genetics and genetic engineering. Not only can humanity's collective genetic inheritance be shaped by selecting which embryos are allowed to develop via pre-implantation genetic diagnosis, but genetic engineering, the availability of the human embryo for experimentation, and combining genes from many species require only sufficient imagination to catalyze the designing of a new humanity.

GENETIC RESOURCES FOR FOOD AND AGRICULTURE

Farm animal genetic resources face a double challenge. On the one hand, the demand for animal products is increasing in developing countries: FAO has estimated that demand for meat will double by 2030; over the same thirty-year period demand for milk will more than double. On the other hand, animal genetic resources are disappearing rapidly worldwide. Over the past 15 years, 300 out of 6 000 breeds identified by FAO have become extinct. Many breeds of local importance for food security are not being improved or utilized in a sustainable manner and are in danger of being lost or diluted by crossbreeding.

Conservation and development of local breeds is important because many of them utilize lower quality feed, are more resilient to climatic stress and to local parasites and diseases, and represent a unique source of genes for improving health and performance traits of industrial breeds. It is also important to develop and utilize local breeds that are already adapted to their environments, most of which are harsh, with very limited natural and managerial input. Animals genetically adapted to these conditions are expected to be more productive at lower costs, support food, agriculture and cultural diversity, and be effective in achieving local food security objectives.

Local communities depend on these adapted genetic resources in many countries. Their disappearance or drastic modification, for example, by crossbreeding, absorption or replacement by exotic breeds, will have serious negative impacts on these human populations. Presently, most breeds at risk of extinction are not supported by any established conservation programmes or active conservation through sustainable utilization (breeding plans) and therefore breed extinction rates are increasing globally.

THE GLOBAL STRATEGY FOR THE MANAGEMENT OF FARM ANIMAL GENETIC RESOURCES

The key component of the *global strategy* is the country-based planning and implementation infrastructure, which includes five structural elements:

- The *global focal point* at FAO headquarters leads the planning, development and implementation of the overall strategy; develops and maintains the information and communication systems; oversees preparation of guidelines; coordinates regional activity; prepares reports and documents for meetings; facilitates policy discussions; identifies training, education and technology transfer needs; develops programme and project proposals; and mobilizes donor resources.
- *Regional focal points* facilitate regional communications; provide technical assistance and leadership; coordinate regional training, research and planning activities; help develop regional policies; assist in identifying project priorities and proposals, and interact with government agencies, donors, research institutions and non-governmental organizations.
- *National focal points* lead, facilitate and coordinate country activities; identify capacity-building needs; develop project proposals; assist with the development and implementation of country policies; and interface with national stakeholders, regional focal points and the global focal point.
- *Donor and stakeholder involvement* is necessary to provide financial and institutional support to the *global strategy.* In this context, the global focal point seeks to ensure stakeholder involvement in all major aspects of the *global strategy,* facilitating opportunities for governmental and non-governmental contributions.
- *DAD-IS, the Domestic Animal Diversity Information System,* is a widely available and easily accessible global database and information source. This global facility makes it possible to share data and information among countries, allowing a rapid and cost-effective

distribution of guidelines, reports and meeting documents, and provides a platform to exchange views and address specific information requests, linking breeders, scientists and policy-makers. A key feature of DAD-IS is the breeds database, which provides the data for the early warning system for animal genetic resources through the World Watch List for Domestic Animal Diversity, whose third edition was released in 2001.

FIRST REPORT ON THE STATE OF THE WORLD'S ANIMAL GENETIC RESOURCES

As part of the *global strategy for the management of farm animal genetic resources,* FAO invited 188 countries to participate in the first *Report on the State of the World's Animal Genetic Resources,* which is to be completed by 2006. To date, 151 countries have accepted to submit country reports. Guidelines for preparation of country reports have been published in Animal Genetic Resources Information Bulletin (FAO) no. 30. These guidelines are used to assist countries in preparing reports as strategic policy documentation covering the state of animal genetic resources, the state of the art and national capacity to manage these resources, and country needs and priorities. Country reports will serve as the base documentation for the State of the World Reporting Process; thus the involvement of all stakeholders in the development of these reports is strongly encouraged.

The objective of the country and global assessments is to provide a comprehensive analysis of the status and trends of the world's farm animal biodiversity and of their underlying causes, as well as of local knowledge regarding its management.

The task is to go beyond description of the resources by analysing the state of these resources and the capacities to manage them, drawing lessons from past experiences and identifying problems and priorities. Country reports are policy documents covering three strategic questions: *Where are we? Where do we need to be? How do we get to where we need to be?* Country reports are intended to be used in planning and implementing priority country actions. In addition, the country report will serve as documentation for the development of the regional and global reports on strategic priorities for action and, subsequently, the first *Global Report on the State of Farm Animal Genetic Resources.*

Country reports provide an assessment in three major areas:

- the *state of diversity* to evaluate the state of conservation, erosion and utilization of farm animal agricultural biodiversity, and an analysis of the underlying processes;
- the *state of national capacity* to manage animal genetic resources, including existing policies, management plans, institutional infrastructures, human resources and equipment;

- the *state of the art* of the available methodologies and technologies to assist farmers, breeders and scientists to better understand, use, develop and conserve animal genetic resources and thereby contribute to global food security and rural development.

International organizations are also being invited to contribute to the state of the world's animal genetic resources preparatory process by providing reports. The long-term aim of the process is for countries and regions to build on the analyses contained in the country reports in order to plan and implement appropriate management of their farm animal genetic resources.

The first *Report on the State of the World's Animal Genetic Resources* will contain the *Report on Strategic Priorities for Action* and will be based on a synthesis of country reports, thematic studies and reports from international organizations.

FIELDWORK AT COUNTRY AND REGIONAL LEVELS

Countries were requested to nominate a national focal point and designate a national coordinator to facilitate the development of the country network on the management of animal genetic resources and to serve as official contact with the global focal point. Keeping in mind that the process involves both scientific and policy matters, the establishment of a National Consultative Committee is recommended to identify the primary areas and issues that need to be addressed in the preparation of the country report and to oversee its preparation. It is essential that the National Consultative Committee have wide and diverse representation and develop a broad network to ensure opportunities for all stakeholders to contribute to the country report.

The response of countries to the invitation of FAO's Director-General to participate in the first *Report on the State of the World's Animal Genetic Resources* and submit a country report has been very positive. During part of 2001 and 2002, FAO trained almost 400 professionals from 178 countries in the preparation of national reports. At the moment, FAO has a team of 15 consultants working in 14 country groupings in all regions of the world. Most countries have undertaken the organization of national stakeholder workshops to elaborate their animal genetic resources policies leading to the country reports.

FAO has organized 14 subregional workshops to discuss draft country reports and regional priorities for action. This has promoted regional cooperation and allows countries that may be experiencing delays to catch up with those in a more advanced state of country report preparation and learn from their experiences. These sessions were coordinated by the regional facilitators acting as FAO consultants.

FAO has provided technical and financial support to 115 countries with contributions from the Governments of the Netherlands and Finland and from the Nordic Gene Bank. FAO and the World Association for Animal Production

(WAAP) signed an agreement to provide technical and operational support for the state of the world animal genetic resources reporting process, including training and country follow-up. FAO considers this cooperation a prime example of effective collaboration with an international non-governmental organization.

GENETIC ENGINEERING

Genetic engineering, also called genetic modification, is the direct manipulation of an organism's genome using biotechnology. New DNA may be inserted in the host genome by first isolating and copying the genetic material of interest using molecular cloning methods to generate a DNA sequence, or by synthesizing the DNA, and then inserting this construct into the host organism. Genes may be removed, or "knocked out", using a nuclease. Gene targeting is a different technique that uses homologous recombination to change an endogenous gene, and can be used to delete a gene, remove exons, add a gene, or introduce point mutations. An organism that is generated through genetic engineering is considered to be a genetically modified organism (GMO). The first GMOs were bacteria in 1973; GM mice were generated in 1974. Insulin-producing bacteria were commercialized in 1982 and genetically modified food has been sold since 1994.

Genetic engineering techniques have been applied in numerous fields including research, agriculture, industrial biotechnology, and medicine. Enzymes used in laundry detergent and medicines such as insulin and human growth hormone are now manufactured in GM cells, experimental GM cell lines and GM animals such as mice or zebrafish are being used for research purposes, and genetically modified crops have been commercialized.

This article focuses on history and methods of genetic engineering, and on applications of genetic engineering and of genetically modified organisms (GMOs). The article on GMOs focuses on what organisms have been genetically engineered and for what purposes. The two articles cover much of the same ground but with different organizations (sorted by application in this article; sorted by organism in the other). There are separate articles on genetically modified crops, genetically modified food, regulation of the release of genetic modified organisms, and controversies.

The DNA contained in genes determines inherited characteristics. Modifying DNA to remove, add, or alter genetic information is called genetic modification or genetic engineering. In the early 1980s, scientists developed recombinant DNA techniques that allowed them to extract DNA from one species and insert it into another. Refinements in these techniques have allowed identification of specific genes within DNA—and the transfer of that particular gene sequence of DNA into another species. For example, the genes responsible for producing insulin in humans have been isolated and inserted into bacteria. The insulin that is then produced by these bacteria, which is identical to human insulin, is then isolated and given to people who have diabetes. Similarly, the genes that produce chymosin, an enzyme that is involved in cheese manufacturing, have also been inserted into bacteria. Now, instead of having to extract chymosin from the stomachs of cows, it is made by bacteria. This type of application of genetic engineering has not been very controversial. However, applications involving the use of plants have been more controversial.

Among the first commercial applications of genetically engineered foods was a tomato in which the gene that produces the enzyme responsible for softening was turned off. The tomato could then be allowed to ripen on the vine without getting too soft to be packed and shipped.

As of 2002, over forty food crops had been modified using recombinant DNA technology, including pesticide-resistant soybeans, virus-resistant squash, frost-resistant strawberries, corn and potatoes containing a natural pesticide, and rice containing beta-carotene. Consumer negativity towards biotechnology is increasing, not only in the United States, but also in the United Kingdom, Japan, Germany, and France, despite increased consumer knowledge of biotechnology. The principle objections to biotechnology and foods produced using genetic modification are: concern about possible harm to human health (such as allergic responses to a "foreign gene"), possible negative impact to the environment, a general unease about the "unnatural" status of biotechnology, and religious concerns about modification.

BIOTECHNOLOGY IN ANIMALS

The most controversial applications of biotechnology involve the use of animals and the transfer of genes from animals to plants. The first animal-based application of biotechnology was the approval of the use of bacterially

Scientists inserted daffodil genes and other genetic material into ordinary rice to make this *golden rice.* The result is a strain of rice that provides vitamin A, a nutrient missing from the diets of many people who depend on rice as a food staple.

CONCERNS ABOUT FOOD PRODUCTION

Some concerns about the use of biotechnology for food production include possible allergic reactions to the transferred protein. For example, if a gene from Brazil nuts that produces an allergen were transferred to soybeans, an individual who is allergic to Brazil nuts might now also be allergic to soybeans. As a result, companies in the United States that develop genetically engineered foods must demonstrate to the U.S. Food and Drug Administration (FDA) that they did not transfer proteins that could result in food allergies. When, in fact, a company attempted to transfer a gene from Brazil nuts to soybeans, the company's tests revealed that they had transferred a gene for an allergen, and work on the project was halted. In 2000 a brand of taco shells was discovered to contain a variety of genetically engineered corn that had been approved by the FDA for use in animal feed, but not for human consumption. Although several anti-biotechnology groups used this situation as an example of potential allergenicity stemming from the use of biotechnology, in this case the protein produced by the genetically modified gene was not an allergen. This incident also demonstrated the difficulties in keeping track of a genetically modified food that looks identical to the unmodified food. Other concerns about the use of recombinant DNA technology include potential losses of biodiversity and negative impacts on other aspects of the environment.

SAFETY AND LABELLING

In the United States, the FDA has ruled that foods produced though biotechnology require the same approval process as all other food, and that there is no inherent health risk in the use of biotechnology to develop plant food products. Therefore, no label is required simply to identify foods as products of biotechnology. Manufacturers bear the burden of proof for the safety of the food. To assist them with this, the FDA developed a decision-tree approach that allows food processors to anticipate safety concerns and know when to consult the FDA for guidance. The decision tree focuses on toxicants that are characteristic of each species involved; the potential for transferring food allergens from one food source to another; the concentration and bioavailability of nutrients in the food; and the safety and nutritional value of newly introduced proteins.

BIOTECHNOLOGY AND GLOBAL HEALTH

The World Health Organization estimates that more than 8 million lives could be saved by 2010 by combating infectious diseases and malnutrition through developments in biotechnology. A study conducted by the Joint Centre for Bioethics at the University of Toronto identified biotechnologies with the greatest potential to improve global health, including the following:

- Hand-held devices to test for infectious diseases including HIV and malaria. Researchers in Latin America have already made breakthroughs with such devices in combating dengue fever.
- Genetically engineered vaccines that are cheaper, safer, and more effective in fighting HIV/AIDS, malaria, tuberculosis, cholera, hepatitis, and other ailments. Edible vaccines could be incorporated into potatoes and other foods.
- Drug delivery alternatives to needle injections, such as inhalable or powdered drugs.
- Genetically modified bacteria and plants to clean up contaminated air, water, and soil.

- Vaccines and microbicides to help prevent sexually transmitted diseases in women.
- Computerized tools to mine genetic data for indications of how to prevent and cure diseases.
- Genetically modified foods with greater nutritional value.

THE PROMISE OF AGRICULTURAL BIOTECHNOLOGY

Due in large part to scientific advances in crop breeding and farming techniques, world food production has doubled since 1960, and productivity from agricultural land and water usage has tripled. Today, the peoples of the world farm an area about the size of South America; without the scientific advances of the past 30 years, farmland equivalent to the entire Western hemisphere would be required.

But a dilemma lies ahead. The world's population is expected to double by the year 2030 to 12 billion, and it is not clear whether current food production is keeping pace with population growth. The question is how best to feed billions of additional people without destroying much of the planet in the process. The disappearance of tropical rain forests, wetlands, and other vital habitats will accelerate unless agriculture somehow becomes more productive and less taxing to the environment.

It seems certain that agricultural biotechnology will play a major role in resolving this dilemma. Biotechnology can be employed to improve the quality of seed grains; increase protein levels in forage crops; and instill in crops resistance to disease, insects, and viruses, as well as tolerance for droughts, floods, and extreme temperatures. In addition, biotechnology can make foods healthier and more nutritious. For example, tomatoes and other fruits and vegetables containing increased levels of certain nutrients, such as vitamins C and E and beta carotene, may help protect against risk of chronic diseases, such as some cancers and heart disease.

Although major genetic improvements have been made in crops, progress in conventional breeding programs has been slow. Moreover, most crops grown in the United States produce less than 50 percent of their genetic potential. These shortfalls in yield are due in large part to the inability of crops to tolerate or adapt to environmental stresses, pests, and disease. For example, some of the world's highest yields of potatoes are in Idaho under irrigation, but in 1993 both quality and yield were reduced severely because of cold, wet weather and widespread frost damage during June. And some of the world's best bread wheats and malting barleys are produced in the north-central states, but in 1993 the disease Fusarium (headblight of wheat and barley) caused an estimated $1 billion in damage. Major advances also have been made through conventional breeding and selection of livestock. Feed efficiency (1) for poultry and swine has been increased by nearly 50 percent.

Milk production per cow has more than doubled since 1955. Diseases such as hog cholera and pests such as screwworm have been eradicated. Yet major diseases of livestock still go uncontrolled and some still go undiagnosed or even unrecognized. These and many other agricultural production hazards have defied traditional solutions such as classical breeding approaches. New research is needed to address these problems, in order to minimize risks; improve the financial stability of farms, rural communities, and agribusinesses; and assure the long-term competitiveness of U.S. agriculture in the world market. These advances must be accompanied by reduced dependency on pesticides and improved environmental sustainability.

Biotechnology can compress the time frame required to translate fundamental discoveries into applications. With improved technology and knowledge about agricultural organisms, processes, and ecosystems, opportunities will emerge to produce new and improved agricultural products in an environmentally sound manner.

This chapter highlights five broad priorities in agricultural biotechnology research that merit attention by Federal agencies:

- Continue mapping and sequencing of animal/plant/microbial genomes to elucidate gene function and regulation and to facilitate the discovery of new genes as a prelude to gene modification.
- Determine biochemical and genetic control mechanisms of metabolic pathways in animals, plants, and microbes that may lead to products with novel food, pharmaceutical, and industrial uses.
- Extend understanding of the biochemical and molecular basis of growth and development including structural biology of plants and animals.
- Elucidate the molecular basis of interactions of plants and animals with their physical and biological environments, as a basis for improving the organisms' health and wellbeing.
- Enhance food safety assurance methodologies, such as rapid tests for identifying chemical and biological contaminants in food and water.

This chapter identifies, within each priority area, selected areas where additional research is needed to assure a steady supply of high-quality agricultural products at reasonable prices for decades to come. The report also highlights recent scientific advances involving Federally supported research.

GENE SEQUENCING AND MAPPING

PRIORITY: Continue mapping and sequencing of animal/plant/microbial genomes to elucidate gene function and regulation, and to facilitate the discovery of new genes as a prelude to gene modification.

To locate desired genetic traits of organisms and effectively use molecular probes to identify DNA sequences, researchers need precise and relatively complete genetic maps. The combination of detailed maps and DNA sequences of genomes can provide a guide to plant and animal breeders for the development of new, efficient, and informed breeding strategies.

The Federally funded Human Genome Project is rapidly advancing gene mapping and sequencing technologies as well as data management techniques for analysing genome data from humans and other organisms. These advances have the potential to make important contributions to the genetic improvement of crops and animals. The agricultural research community is in an ideal position to apply the technical and scientific advances from the Human Genome Project to address the biology of organisms important to agriculture, aquaculture, and forestry.

PLANT GENOME RESEARCH

The U.S. agricultural research community has mounted a considerable effort to understand the structure, function, and regulation of genes in a wide range of crop and forestry species. There may never be resources sufficient to sequence completely the genomes of all agriculturally important plants, but there is a high degree of similarity among species. Therefore, it is cost-effective to select one representative plant for concerted sequencing of its entire genome.

The model experimental plant *Arabidopsis thaliana* offers the best chance for achieving a completely sequenced plant genome in the foreseeable future.

Arabidopsis has the smallest known genome of any flowering plant, with 100 million base pairs and little repetition in its DNA.

Detailed genetic and physical maps have been developed, and understanding of the plant's biology is growing rapidly. A complete understanding of the *Arabidopsis* genome would offer enormous potential for improving agriculture. In addition to studies on *Arabidopsis*, parallel studies will be required at some level with plants used to grow food and fibre.

Knowledge of a single gene can have broad impact. An example is a recent breakthrough in the identification and sequencing of a tomato gene that confers resistance to a bacterial disease. The sequencing of this gene has given scientists the first unambiguous glimpse of the resistance response of a plant to one of its pathogens. This discovery has opened the way to identification of disease resistance genes in other plants as well as created an avenue for investigating the molecular basis for disease resistance.

GENETICS OF RHIZOSPHERE MICROORGANISMS

Most farms grow only three or four crops, and some grow the same crops year after year in the same fields. These practices promote soil infestations by pathogens that damage or destroy roots.

For most crops, soilborne plant pathogens generally are uncontrolled, except by limited crop rotations and some host plant resistance.

Recent studies show that some soil microorganisms associated with the roots of crops carry and express genes for defence of roots against pathogens. Little is known about these beneficial microorganisms, which represent an enormous untapped genetic resource for improving crop production and efficiency of fertilizer use. Identification and genetic studies of these microorganisms could enable their direct deployment or use of their genes in the development of disease-resistant crops.

NEW USES FOR PLANT MATERIALS

Higher plants are very efficient sources of renewable organic materials. In particular, corn starch, which is both a useful raw material and an important human and animal food, is produced in large quantities at about the same cost per pound as is crude petroleum. Many genes involved in starch biosynthesis have been cloned recently and it should be possible to explore the development of new kinds of starches for use in nonfood applications. The potential use of starch as a co-polymer in biodegradable plastics is being explored.

Plant storage lipids (*i.e.,* oils) can be produced in large quantities, and they represent a chemically versatile form of biomass. Enhancement of oil yield and composition in plants such as palm and rapeseed depends on increased understanding of the factors that regulate lipid biosynthesis. Progress has been made in identifying genes and enzymes involved in the early steps of fatty acid and lipid biosynthesis. To date, the most promising approach to manipulating oil yield and composition involves modifying the vast variety of fatty acids found as lipid constituents in various wild species. This is technically possible only if the basic biochemistry is elucidated.

NOVEL PRODUCTS FROM PLANTS

Higher plants synthesize an enormous spectrum of chemical constituents. Many have known value as drugs, biomaterials, solvents, flavorings, fragrances, or coloring agents. Through development of transgenic plants, the purity of many plant-derived chemicals can be enhanced and the range of chemicals expanded. One example is the pending development of plants that produce hydroxylated fatty acids. These oils can be used in applications ranging from hydraulic fluids to nylon synthesis. Similarly, the recent development of plants that accumulate biodegradable thermoplastics illustrates the potential for producing completely new compounds in plants.

Many natural chemicals are important in plant defence mechanisms or in abiotic stress responses. Some chemicals confer important horticultural qualities to plants. Thorough studies have been conducted on the biosynthesis of certain classes of these compounds, such as flavonoids, cyanogenic glycosides, and certain alkaloids. Other classes of compounds, such as terpenoids, have received relatively little attention. Terpenoids are important for plant growth and can be used as essential oils, resin acids, and pigments. An efficient mechanism must be developed to obtain specific biochemical information about these classes of compounds, to complement the extensive ongoing efforts to identify them and elucidate their structures. Biotechnology offers promise for accelerating the characterization of these substances as well as facilitating their production for commercial use.

GROWTH AND DEVELOPMENT

Priority: Extend understanding of the biochemical and molecular basis of growth and development including structural biology of plants and animals.

Biotechnology offers opportunities to study a wide variety of genetically regulated growth and development processes. Alteration of these processes can help improve product quality and nutritional and economic benefits. The genetic makeup of plants and animals can be altered by either insertion of new, useful genes or removal of unwanted ones. Discoveries made through biotechnological approaches are changing the way plants and animals are grown, boosting their value to growers, processors, and consumers alike.

The many opportunities in plant research. The numerous potential benefits range from the development of specialty starch polymers for packaging and oils for use as industrial lubricants to modification of post-harvest traits such as ripening and senescence. The recent development of tomatoes with modified ripening characteristics may lead to extended post-harvest shelf life of fruits, vegetables, and flowers. Information about the biological basis of cold tolerance may result in the development of cold-hardy flowers and ornamentals now grown only in subtropical or tropical climates. The engineering of increased salt tolerance in plants may offset losses due to saline soils.

PLANT FLOWERING

Scientists are gaining their first glimpse into the complex mechanisms of flowering. Most plants use environmental cues to initiate flowering so that individuals of the same species flower synchronously, thus maximizing the chances of successful outcrossing. Scientists have discovered molecules in plants called phytochromes that sense dark and light conditions and send signals that regulate plant activities such as sprouting, growth, and flowering.

There is still much to learn about how flowers develop. Investigations of genes controlling floral meristems — the tissue that forms flowers — promise to help explain the genetic mechanisms that manage the establishment and maintenance of this tissue. Ongoing research also is extending understanding of the physiological processes that take place during the transition from vegetative to reproductive development.

MECHANISMS OF PLANT FERTILIZATION

From apples to zucchini, most foods originate from the fertilized flowers of plants. In tomatoes, for example, several genes have been discovered that act within a flower's anthers, where pollen is produced. Promoters (regions of genes that act as on-off switches) also have been identified. Scientists now are pinpointing the overall molecular mechanisms that cue tomato pollen to fertilize the female part of a tomato flower.

Increased understanding of plant fertilization eventually could lead to incremental improvements in the yield and energy efficiency of crop production.

Some plants have genes that render their anthers infertile. This male sterility allows production of higher yielding hybrid seeds, whereby male-sterile plants are fertilized with pollen from specially selected sources. Genetically engineered systems for controlling male fertility are used to produce hybrid seed of important crops such as oilseed rape, corn, and rice. The technology also is applicable to tomato, lettuce, and a wide range of other crop plants.

These advances eliminate the need for costly, labour-intensive hand or mechanical removal of anthers in making hybrid crosses and will have a major impact on the billion-dollar hybrid seed industry. In addition, the knowledge gained by studying anther development and fertilization at the molecular and genetic levels will reveal other approaches that can be used to produce novel varieties of hybrid crop plants.

BIOLOGICAL BASIS OF PLANT FORM

Conventional breeding often is aimed at achieving certain visual traits. Once the biological basis for these traits is understood, biotechnology could provide the tools to change plant architecture precisely and optimize plant form. For example, changes in leaf form and/or number could maximize photosynthetic capacity and increase production, and changes in root architecture could facilitate and maximize water and mineral capture and uptake. Also, the tools of biotechnology could be exploited to design crops that thrive on soils contaminated with heavy metals, such as cadmium or mercury. Scientists are studying and beginning to redesign the genes responsible for the activity of phytochelatins, natural compounds in plants that bind these metals and detoxify them.

ANIMAL GROWTH AND DEVELOPMENT

Research also is needed to explore the genetic basis of animal growth and development. Technologies have been developed for gene cloning, gene transfer, *in vitro* culturing, and sex determination of embryos, and these approaches are being refined for use with animals. Transgenic embryos or offspring have been produced from rabbits, chicken, fish, sheep, swine, and cattle. Genes can be targeted for expression in specific tissues, but the efficiency of gene transfer methods should be improved, and genes important for growth and development must be characterized further.

ENVIRONMENTAL INTERACTIONS

Priority: Elucidate the molecular basis of interactions of plants and animals with their physical and biological environments, as a basis for improving the organisms' health and wellbeing. Future gains in agricultural productivity and sustainability depend heavily on the use of biotechnology to improve the health and wellbeing of agriculturally important plants and animals.

Research on environmental interactions is emerging rapidly as a powerful means of increasing resistance to stress and disease, with consequent economic and environmental benefits. For example, basic research on the genetic control of biochemical responses to water stress in drought-hardy native plants can suggest how crop plants might be modified genetically for more consistent performance in years of drought. Similar studies are needed to improve the tolerance of U.S. crops to heat stress, winter and frost damage, harm caused by salinity, and other physical stresses.

New crop genotypes also must be resistant to an expanding diversity of pests and diseases. Basic research at the organismal and cellular levels, together with classical breeding techniques, already has paid significant dividends to farmers, consumers, and the environment. Today, U.S. wheat, corn, soybean, barley, and sorghum are grown without fungicides because resistant varieties have been bred. However, these and most other U.S. crops still are subject to major damage from insects and diseases caused by soilborne plant pathogens and insect-vectored viruses, for which there has been no useful source of resistance. As a result, U.S. agriculture continues to depend on pesticides. The survival of high-value horticultural crops depends largely on soil fumigants.

Biotechnology also can help reduce the billions of dollars in agricultural losses caused by animal diseases. There are opportunities to boost the immune competency of animal hosts by genetic manipulation, develop biotechnology-derived regulators of immune function, and identify ways to prevent attachment of disease agents to host cells. Factors affecting the disease-producing capacity of a bacterium or virus can be identified, and strategies can be devised to help the host withstand these pathogens. Through the use of techniques such as enzyme-linked immunosorbent assays, polymerase chain reaction (PCR), and monoclonal antibody-based systems, biotechnology can contribute to major advances in diagnostics.

PLANT PESTS AND DISEASES

Virologists have transferred virus genes — such as those for production of virus coat proteins — to plants, thereby conferring resistance to those viruses in otherwise susceptible hosts. Some genes also have been shown to limit virus replication and associated damage.

Replication depends on the regulatory and metabolic processes of the host; these processes must be understood before scientists can engineer virus-resistant plant varieties. Another possible way to increase plant resistance to viral infections would be to interfere with the spread of the virus, but this approach would require an understanding of the biochemical mechanisms of spread and transmission.

All crop plants are susceptible to bacterial diseases, which have been difficult to control through plant breeding. Bactericides are not a complete solution, either, because bacteria quickly evolve resistance to them. The bacterial disease fire blight is destructive to pears, apples, quince, and some ornamental plants. This disease has plagued fruit producers for over three centuries, but its mechanism of attack only recently has been revealed using the tools of molecular biology. This knowledge is expected to provide ways to combat this disease. Nearly all agriculturally important fruit, vegetable, and grain varieties acquire fungal diseases that can cost growers billions of dollars annually. Information is accumulating rapidly on the molecular basis of plant responses to infection by these pathogens. In many cases, the difference between resistance and susceptibility is the rate of response; if rapid, then the plant is resistant. Scientists have found that deliberately disarmed fungal pathogens can trigger the necessary response in susceptible (*i.e.*, slow responding) plants before the virulent pathogen arrives. This approach has opened a new avenue in disease control: the use of "pathogen derived" biocontrol agents to induce resistance in otherwise susceptible plants. This technique is only beginning to emerge as a way to control heretofore unmanageable plant diseases.

CONTROL OF INSECT PESTS PREYING ON PLANTS

Insect pests cause major crop damage. Examples include *Phylloxera*, which has caused devastation in European and Californian vineyards, and *Psylla*, which has ravaged the apple and pear industry of the eastern United States. In this century, the United States has imported and released approximately 800 natural enemies of insect pests; about 40 percent are providing some level of biological control. However, this approach to insect pest management has its limits. Further progress will require increased knowledge of interactions between pests and their natural enemies, and, if necessary, appropriate genetic modifications.

The chemicals emitted by plants in response to feeding by insect pests serve as an attractant to the natural enemies of some insect pests. Corn leaves attacked by beet armyworm caterpillars release volatile compounds that attract wasps that are parasites of these armyworms. Moreover, these parasitic wasps can recognize and fly towards the source of volatile compounds produced in response to feeding by their caterpillar hosts; the wasps ignore the odours released from leaves damaged mechanically, such as by mowing. Expanded knowledge of the "information molecules" used by beneficial insects to find their prey on crops could lead to development of crops capable of producing stronger signals, to attract higher populations of beneficial insects. To exploit fully this approach to biological control, more research is needed on the biochemistry and molecular biology of insect pests and their natural enemies.

Due to insects' evolving resistance to pesticides, and the removal of many of these chemicals from the market, there are few effective means of controlling sucking insects, such as whiteflies, aphids, and leafhoppers. These same insects are among the most important vectors (carriers) of viruses, rickettsia-like bacteria, and mycoplasma-like organisms responsible for major plant diseases. New synthetic chemical compounds to be on the market in the next few years primarily target chewing insects. Biological methods to control sucking insects also are advancing at the molecular level. An example is the enzymatic preparation of synthetic pyrethroids.

The salivary glands of aphids, whiteflies, and leafhoppers contain bacterial symbionts that provide essential amino acids to their insect hosts. Only a few of these symbionts have been characterized. Research is needed on the basic biology, molecular biology, and production of these bacterial symbionts. If the bacterial symbionts could be disrupted in some way, either by manipulating the genes of the symbionts or by engineering an antimicrobial agent into a plant, then an innovative control mechanism could be developed.

BENEFICIAL MICROORGANISMS IN PLANTS

Plants selectively aid the growth of specific types of beneficial microorganisms. Some microorganisms, for instance, have been shown to provide growth factors for plants and protect plants against insect attack and infection. Information about the molecular signals and genetic controls of these associations can lead to the use of these microorganisms to improve the performance — and especially the consistency of performance— of important food and fibre crops, and even to enhance the effectiveness of the microbes themselves.

ANIMAL PESTS AND DISEASES

Work is underway to introduce new genes into animals in order to impart resistance to disease and to identify host cell traits that can be used to enhance resistance to disease agents. For instance, a specific receptor on pig intestinal cells is required for *E. coli* to bind to the cell.

Strains that do not carry the receptor— or have the receptor deleted through the tools of biotechnology — are less susceptible to the diarrheal disease caused by the bacterium. Other research on enteric rotavirus infections in pigs has identified specific cell membrane components required for attachment of the rotavirus to an intestinal cell. A complementary molecule can be constructed using biotechnological methods to saturate and block this receptor site.

Porcine stress syndrome is an inherited condition in swine that can lead to meat that is of poor quality or unusable. This condition causes significant losses in the swine industry. At the same time, animals susceptible to this genetic disease have highly favourable muscling with low fat content. A candidate gene for this disease has been identified and research now is needed to develop a DNA probe for the trait, as a means to identify susceptible animals and eliminate them from breeding programs. This research will allow the development of swine with high-quality, low-fat meat but without the trait associated with the disease.

FOOD SAFETY

Enhance food safety assurance methodologies, such as rapid tests for identifying chemical and biological contaminants in food and water.

Food-and waterborne illnesses caused by microorganisms pathogenic to humans have been estimated to affect more than 80 million Americans and cost the U.S. economy over $40 billion annually. For example, some 9,000 Americans die each year from food-borne illnesses caused by microorganisms such as *E. coli* in meat and *Salmonella* in poultry. At present, virtually all food inspection is visual. Rapid, accurate, non-invasive tools of biotechnology can help improve the detection and control of food-borne human pathogens as well as chemical contaminants. Federal support is needed in this area to assure that these tools are developed in a timely manner.

DNA PROBES

DNA probe kits have been developed for *Salmonella, Listeria, E.coli* 0157:H7, and *Staphylococcus aureus*. Compared to traditional culture-plating methods, these new diagnostic kits offer greater precision, shorter turnaround times, and reduced need for highly trained personnel.

DNA diagnostic techniques for Norwalk viruses also are commercially available. Culture techniques have not been successful in detecting these leading causes of gastroenteritis, making the availability of genetically based techniques critical to the detection and identification of the Norwalk viruses.

In addition to detecting food contaminants, DNA probes and other tools of biotechnology can help reduce levels of naturally occurring toxicants in foods. DNA probes can be exploited in research and plant breeding to isolate genes associated with the biosynthesis of major toxicants, facilitate understanding of the genetic regulatory mechanisms for toxicants in plants, and develop lines of plants with reduced levels of toxicants.

Mycotoxins in food are a periodic threat to food safety. DNA probes could help detect the presence of mycotoxin-producing fungi that grow under certain conditions in plant materials such as improperly dried corn and peanuts. DNA probes could also be used to learn more about the sources of fungal contamination in the environment and as a means to develop management strategies under field conditions.

BIOSENSORS

The development of biosensors offers great promise for improving food processing, analysis, and safety assurance. The highly specific actions of biological molecules can be exploited for use in biosensors that can measure the concentration of specific components in complex mixtures. Enzymes, antibodies, and microbial cells can be immobilized on solid surfaces, and the specific reactions they mediate can be detected by various physical and chemical means. Biosensors are commercially available to detect a variety of sugars, alcohols, esters, peptides, amino acids, cell types, and antibiotics. Development of tailor-made membranes capable of separating molecules based on size, electrical charge, or solubility will accelerate biosensor development as well as the exploitation of biomimetic systems.

Miniaturization and mass production of biosensors could increase their availability and decrease their unit cost. Technologies such as microlithography, ultrathin membranes, and molecular self-assembly have the potential to facilitate the miniaturization of molecular and cellular processes for enrichment, detection, and analysis of chemical and microbiological contaminants in food. Advances in the semiconductor industry have made it possible to combine chemical and biological components and integrated circuits in miniaturized systems. Biosensors can be inserted directly into food processing streams to obtain on-line, real-time measurements of important food processing parameters. Miniature biosensors also could be incorporated into food packages to monitor temperature stress, microbial contamination, or remaining shelf life, and to provide a visual indicator to consumers of product state at the time of purchase.

SYSTEMATIC STUDY OF FOOD-BORNE MICROBES

An appropriately designed long-term molecular biology study could characterize and correlate food-associated microbiological isolates in existing culture collections. Food-related isolates from these collections could be used to screen foods for microbiological contamination. Foods identified in this way then would serve as focal points for regulatory agencies and critical systems aimed at improving control over growing, shipping, processing, distribution, and other steps during which these foods may become contaminated.

MARINE BIOCHEMICAL PROCESSES AND BIOMOLECULAR MATERIALS

The Marine Processes objective involves the description of the dynamics of coastal waters off Southern Iberia, combined with chemical characterization, to accurately assess the primary productivity in shelf waters. CIMA's approach to these phenomena is based on modelling, analyses and observations both in situ and remote. The major topics covered are:

- Models and measurements of transfer phenomena in the ocean,
- Chemical dynamics in the ocean,
- Marine microbial dynamics and foodwebs.

Recent research has demonstrated that marine biochemical processes can be exploited to produce new biomaterials. For example, a corporation in Chicago is commercializing a new class of biodegradable polymers modelled on natural substances that form the organic matrices of mollusk shells. Equally exciting are the mechanisms used by marine diatoms, coccolithophorids, mollusks, and other marine invertebrates to generate elaborate mineralized structures on a nanometer scale (less than a billionth of a meter in size). Nanometer-scale structures can have unusual and useful properties.

Research that will enhance understanding and allow engineering of the processes for creating these bioceramics promises to revolutionize the manufacturing of medical implants, automotive parts, electronic devices, protective coatings, and other novel products. Biomaterials also hold promise for counteracting biofouling, which long has been recognized as an extensive and costly problem. Bacterial biofilms form slime layers that increase drag on moving ships, interfere with transfer on heat exchangers, block pipelines, and contribute to corrosion on metal surfaces. Bacterial and microalgal colonization of surfaces is accompanied by settlement of invertebrate larvae and algal spores, eventually leading to "hard fouling" and the need for costly cleaning.

The most effective anti-fouling coatings have utilized toxic chemicals, such as copper and organotins. There is an urgent need for nontoxic biofouling control strategies, due to heightened recognition of the impact that toxic coatings can have on the environment. Research is needed on the attachment mechanisms of marine organisms and the natural products they employ to prevent fouling of their own surfaces. Molecular approaches to characterizing biofilm structure and development offer considerable potential for finding novel biofouling prevention strategies. It is now possible to determine the genes and pathways involved in regulation and synthesis of bacterial adhesive polymers.

Considerable progress has been made in understanding the nature and expression of surface polymers produced by microorganisms such as the nitrogen-fixing *Rhizobium* species and the opportunistic pathogen *Pseudomonas aeruginosa*. Similar approaches can been applied to marine biofilm bacteria, to find the genetic determinants of adhesive production and the environmental factors that regulate synthesis.

Molecular biology techniques also can be applied to determine the basis of natural antifouling mechanisms. Many marine plants and animals remain free of attached bacteria, either because they produce repelling compounds or because their surface structure neutralizes bacterial adhesives. A product generated by the seagrass *Zostra marina* (eelgrass), for example, is an effective agent for preventing fouling by bacteria, algal spores, and a variety of hard-fouling barnacles and tube worms. Molecular characterization of natural fouling resistance could provide new strategies for fouling control. Potential applications include prevention of fouling in industrial pipelines or heat exchangers, improved design of trickling filters or aquaculture circulation systems, and control of biofilm infections of medical implants and prosthetic devices.

BIOMONITORS

Marine organisms can provide the basis for development of biosensors, bio-indicators, and diagnostic devices for medicine, aquaculture, and environmental monitoring. One type of biosensor employs the enzymes responsible for bioluminescence. The *lux* genes, which encode these enzymes, have been cloned from marine bacteria such as *Vibrio fischeri* and transferred successfully to a variety of plants and other bacteria. The *lux* genes typically are inserted into a gene sequence, or operon, that is functional only when stimulated by a defined environmental feature.

The enzymes responsible for toluene degradation, for example, are synthesized only in the presence of toluene. When *lux* genes are inserted into a toluene operon, the engineered bacterium glows yellow-green in the presence of toluene. This genetically engineered system "reports" that biodegradation of a specific chemical, in this case toluene, is proceeding.

Another type of biomonitor that holds great promise is the gene probe, which can be used to identify organisms that pose health hazards or may be useful in research.

Specific gene probes can be employed, for example, to detect human pathogens in seafood and recreational waters; fish pathogens in aquaculture systems; microorganisms capable of mediating desired chemical transformations (*e.g.*, toxic chemical degradation, CO_2 assimilation, metal reduction); and specific fish stocks in fish migration and recruitment studies.

BIOPESTICIDES

Natural marine products have the potential to replace chemical pesticides and other agents used to maximize crop yields and growth.

Continued Federal support for R&D in this area is likely to result in useful natural pesticides that would provide greater specificity and fewer harmful side effects than do conventional synthetic agents. Current U.S. expenditures for all pesticides amount to $47 billion annually; by the year 2000, biopesticides from marine and other sources are expected to capture an estimated 10 percent of this market.

An example of a marine biopesticide in use today is Padan*TM*, which was developed from a bait worm's toxin known to ancient Japanese fishermen. This natural pesticide has demonstrated activity against larvae of the rice stem borer, the rice plant skipper, and the citrus leaf miner, among other pests. More recently, scientists in Montana discovered novel compounds in marine algae and marine sponges containing symbiotic microorganisms. These compounds promoted growth and stimulated germination and increased root and coleoptile lengths in test plants.

Several sponge and nudibranch species produce terpenes, a broad class of aromatic compounds used in solvents and perfumes and known to deter feeding by fish. Extracts derived from these same sponge and nudibranch species also demonstrated powerful insecticidal activity against two species, grasshoppers and the tobacco hornworm.

BIOMASS FOR ENERGY PRODUCTION

Approximately 40 percent of all primary energy production, or photosynthesis, occurs in the seas. In this process, oceanic plants (phytoplankton, seaweeds, seagrasses) take up carbon dioxide (CO_2) and, with light energy from the sun, convert it into organic carbon (primarily sugars) and oxygen. The oceans contain 50 times as much carbon dioxide as does the atmosphere, and it is estimated that primary production incorporates 35 gigatons (1 gigaton = 1 x 10*15* grams) of carbon into marine biomass annually. This abundant source of fuel for energy production has not been tapped commercially because it is not competitive with soybean meal and other easily harvested, traditional sources of biomass, and also because, regardless of the source, biomass is not competitive with other types of fuels.

The Federal Government should continue to support research on the use of biotechnology to enhance biomass production and utility. At least three general approaches are being explored.

First, the enzyme that captures CO_2 for photosynthesis — ribulose bisphosphate carboxylase\oxygenase or "RUBISCO" — is relatively inefficient, so supercomputers are being used to verify structural information, and the enzyme is being redesigned to optimize its function. Second, the chemical composition of biomass can be altered to make it more suitable for particular applications.

For example, marine microalgae are being genetically engineered to boost their lipid content, with the aim of providing a source of alternative fuels that is more economical than are conventional sources. Third, biotechnology is being used to convert biomass to ethanol and other alternative forms of energy and chemical feedstocks.

NEW AND IMPROVED PROCESSES FROM THE SEAS

Priority: Develop bioremediation strategies for application in the world's coastal oceans, where multiple uses — including wastewater disposal, recreation, fishing, and aquaculture — demand prevention and remediation of pollution; and develop bioprocessing strategies for improving sustainable industrial processes.

BIOREMEDIATION

Bioremediation shows great promise for addressing problems in marine environments and in aquaculture. These problems include catastrophic spills of oil in harbours and shipping lanes and around oil platforms; movement of toxic chemicals from land, through estuaries, into the coastal oceans; disposal of sewage sludge, bilge waste, and chemical process wastes; reclamation of minerals, such as manganese; and management of aquaculture and seafood processing waste.

The full potential for marine organisms and processes to contribute new waste treatment and site remediation technologies cannot be realized without enhanced understanding of the unique conditions in marine environments. For example, oxidation-reduction (redox) states can fluctuate in coastal and estuarine sediments.

The impact of changing redox conditions on biodegradation of environmental contaminants must be understood before waste management and remediation strategies and predictive models can be developed for contaminated sediments.

BIOPROCESSING

The emerging discipline of bioprocess engineering involves the application of biological science in manufacturing, to produce products such as biopharmaceuticals and natural bioactive agents. Bioprocess engineering requires an understanding of the biological system employed (such as a marine organism), isolation and purification of a product, and translation of the product into a stable, efficacious, and convenient form.

An emerging area of interest is the potential of marine bacteria and fungi to produce unusual chemical structures with no parallels in terrestrial organisms. Small-scale studies have begun to indicate the richness of marine microorganisms as sources for novel lead structures.

AQUACULTURE

Priority: Use the tools of modern biotechnology to improve the health, reproduction, development, growth, and overall wellbeing of cultivated aquatic organisms; and promote the interdisciplinary development of environmentally sensitive, sustainable systems that will enable significant commercialization of aquaculture.

Aquaculture, which long has been practiced in Asia and is increasingly popular in the United States, Europe, and South America, will benefit tremendously from the use of new molecular tools and processes. With worldwide seafood demand projected to increase 70 percent in the next 35 years, and harvests from capture fisheries stable or declining, aquaculture will have to produce seven times as much seafood as it generates now to supply global demand by 2025.

The use of modern biotechnology to intervene in the rearing process and enhance production of aquatic species holds great potential not only to meet this demand, but also to improve U.S. competitiveness in aquaculture. The U.S. aquaculture industry has grown rapidly in recent years. Farm gate receipts exceeded $800 million in 1992 — a fourfold increase since the early 1980s.

The growth and international competitiveness of the U.S. aquaculture industry will be determined by the size of the resource investment in research and technology development. This investment should be made through a partnership of Federal and state agencies and the private sector. The Federal role is to provide leadership in supporting research to advance knowledge in important research areas and to facilitate the transfer of promising results and technologies to the private sector.

The major research issues in aquaculture are similar to those for other agricultural sectors, but the knowledge base for aquaculture is comparatively meager. Development of this knowledge is a particular challenge due to the diversity of cultured aquatic species and the systems for their production. Federal support for biotechnology research in this area will expand the knowledge base and yield significant dividends.

The application of biotechnology promises significant benefits to both producers and consumers of aquacultural products. The use of genetically enhanced organisms may improve production efficiency through improvements in growth rates, food conversion, disease resistance, and product quality and composition. The application of biotechnology to aquaculture also may help conserve wild species and genetic resources and provide unique models for biomedical research.

ENHANCING REPRODUCTION AND EARLY DEVELOPMENT

Biotechnology can be applied to enhance reproduction and early development of cultivated aquatic organisms. The resulting benefits could include year-round production of gametes and fry of economically valuable species and creation of new markets for specialized, genetically improved broodstock. Similarly, biotechnology may provide techniques for improving the reproductive success and survival of endangered species, thereby helping to preserve the diversity of life on Earth. As a first step, research should be directed towards improving basic understanding of environmental, hormonal, biochemical, and genetic control of reproduction.

More specifically, scientists must identify and understand the mechanisms of expression of genes involved in reproduction and development, improve technologies for cryo-preserving gametes and embryos, improve delivery systems for administration of natural and synthetic hormones, and enhance understanding of the pharmacokinetics of uptake and release of administered hormones.

IMPROVING HEALTH AND WELL-BEING

Biotechnology offers substantial opportunities to improve the health and wellbeing of cultivated aquatic organisms. More than 50 diseases affect fish and shellfish cultured in the United States, causing losses of tens of millions of dollars annually. Biotechnology not only can improve the survival, growth, vigour, and wellbeing of cultivated stocks, but also can reduce disease transfer between cultivated and wild stocks. New products and market opportunities can be developed related to aquatic animal health and wellbeing.

The tools of molecular biology can provide a basic understanding of host immunity, resistance, and susceptibility to diseases and associated pathogens by furnishing information about life cycles and mechanisms of pathogenesis, antibiotic resistance, and disease transmission. Improved technologies must be developed for detecting and diagnosing pathogens and diseases and for enhancing the genetic basis of disease resistance, thereby reducing the need for antibiotics and other drugs.

Potential products resulting from this research include gene therapy techniques; broodstock free of pathogens; safe, effective prophylactic agents, including immune modulators, antigens, and vaccines; safe, effective therapeutic agents; and improved systems for administering prophylactic and therapeutic agents.

IMPROVING QUALITY AND VALUE

The Federal Government has a responsibility to help ensure the safety and quality of food supplies, and biotechnology can and should be an invaluable tool in carrying out this mandate. Biotechnology can be employed to assess and improve the safety, freshness, colour, flavour, texture, taste, nutritional characteristics, and shelf life of aquacultural food products. In addition, practical technologies can be developed to detect and assay toxins, contaminants, and residues in seafood, and to reduce or eliminate contaminants.

There are also opportunities to apply biotechnology in improving seafood processing. Research should be conducted to develop and improve technologies for all these applications.

CONSERVING GENETIC RESOURCES

The preservation and enhancement of biodiversity in natural systems is an important Federal priority. Therefore, the Federal Government should encourage and support programs to maintain and enhance biodiversity in aquatic systems through cultivation and stocking of aquatic species. Biotechnology can be employed in two ways to conserve genetic resources of aquatic species.

First, the tools of biotechnology can be used to identify and characterize important aquatic germplasm, including endangered species. Genomes of aquatic species can be analysed and characterized, and quantitative trait loci identified. Second, biotechnology can be applied to improve understanding of the molecular basis of gene regulation and expression as well as sex determination and thereby improve methods for defining species, stocks, and populations. Approaches include developing marker-assisted selection technologies, improving precision and efficiency of transgenic techniques, and improving technologies for the cryo-preservation of gametes and embryos. Ultimately, stocking certain areas with selected, cultivated species and strains could help maintain biodiversity in natural aquatic ecosystems.

ENHANCING BIOMEDICAL MODELS

Aquaculture has important purposes other than food production. Because they often adapt to extreme environments, marine organisms can provide unique models for research on biological and physiological processes.

Studies of the developmental, cellular, and molecular aspects of marine organisms as model systems will provide insights into the basis of disease mechanisms and pathogenesis in humans. By contrast, use of mammalian organisms as a basis for the development of some types of human disease models may be neither feasible nor cost effective.

Progress in this arena will require that sophisticated molecular biology technologies be adapted to marine organisms, in order to enhance understanding of their biological processes. For example, approaches for gene transfer into eggs have been developed for many terrestrial organisms, but not for most marine species.

This technology is needed for analyses of gene regulatory systems and gene expression. In addition, methods need to be developed for culturing tissues from marine organisms. Cultured cell lines will provide opportunities for gene transfer and gene expression studies and enhance the usefulness of marine species as biomedical research models. This is an important research area deserving of Federal support.

UNDERSTANDING AND CONSERVING THE SEAS

Priority: Improve understanding of microbial physiology, genetics, biochemistry, and ecology in order to provide model systems for research and production systems for commerce, and to contribute to understanding and conservation of the seas.

Scientists have a powerful new array of sampling devices and measuring instruments that will accelerate greatly the acquisition of knowledge about ocean resources and foster their wise use.

These technologies include manned deep-sea submersibles, remotely operated vehicles, geosynchronous satellites, sophisticated acoustic measuring devices, pressure-retaining deep-sea samplers, geographic information systems, real-time flow cytometry, PCR and biomonitoring techniques, computerized databases, and other forms of information exchange and analysis.

These tools should be exploited to accelerate the discovery of unknown marine microorganisms and to expand understanding of known varieties.

Federal support for this research is essential, because only then will sufficient information be acquired to assure that practical applications will result. As new life forms and processes become known, and as understanding of them grows, marine biotechnology will make significant contributions to the nation's social and economic wellbeing.

HARMFUL EFFECTS OF GENETICALLY MODIFIED FOODS

Largely between 1997 and 1999, genetically modified (GM) food ingredients suddenly appeared in *2/3rds of all US processed foods*. This food alteration was fuelled by a single Supreme Court ruling. It allowed, for the first time, the patenting of life forms for commercialization. Since then thousands of applications for experimental genetically-modified (GM) organisms, including quite bizarre GMOs, have been filed with the US Patent Office alone, and many more abroad. Furthermore an economic war broke out to own equity in firms that legally claimed such patent rights or the means to control not only genetically modified organisms but vast reaches of human food supplies. This has been the behind-the-scenes and key factor for some of the largest and rapid agri-chemical firm mergers in history.

The merger of Pioneer Hi-Bed and Dupont (1997), Novartis AG and AstraZeneca PLC (2000), plus Dow's merger with Rohm and Haas (2001) are three prominent examples, Few consumers are aware this has been going on and is ever continuing. Yet if you recently ate soya sauce in a Chinese restaurant, munched popcorn in a movie theatre, or indulged in an occasional candy bar-you've undoubtedly ingested this new type of food. You may have, at the time, known exactly how much salt, fat and carbohydrates were in each of these foods because regulations mandate their labelling for dietary purposes. But you would not know if the bulk of these foods, and literally every cell had been genetically altered!

In just those three years, as much as 1/4 of all American agricultural lands or 70-80 million acres were quickly converted to raise genetically-modified (GM) food and crops. And in the race to increase GM crop production verses organics, the former is winning.

CORE PHILOSOPHICAL ISSUES

When Gandhi confronted British rule and Martin Luther King addressed those who disenfranchised Afro-Americans, each brought forth issues of morality and spirituality. They both challenged others to live up to the highest principles of humanity. With the issue of GM food technology, we should naturally do the same, and with great respect for both sides. It is not enough to list fifty or more harmful effects but we need to also address moral, spiritual and especially worldview issues. Here the stakes are incredibly huge. The philosophical issues involving GMOs, why this technology represents the impregnation of a mechanical worldview, a death-centred vision of nature that is greatlyt accelerating the death of species on earth.

FROM HYBRIDIZATION TO GMOS

Another challenging phenomenon to face in our modern world is that of hybridization. It seems to have worked so very successfully in some commercial realms, and as a major application of Gregor Mendel's revolutionary Gene Theory. Mendel offered a logical extension of the larger mechanical worldview. Just as we create factory assembly lines for manufacturing inanimate products, why can't we also manufacture living organisms, and using the same or similar principles? Why not take this assembly-line process to the next logical and progressive level?

What's wrong then with the "advance" of genetic engineering? No doubt, with hybridizations conscious life is manipulated. But living organisms continue to make some primary genetic decisions amid limited selections. We can understand this with an analogy. There is an immense difference between being a matchmaker and inviting two people to a dinner party, to meet and see if they are compatible. This differs essentially from forcing their meeting and union or a violent date rape. The former act may be divine, and the latter considered criminal. The implication is that biotechnology involves vital moral issues in regard to the whole of life in nature.

With biotechnology, roses are no longer crossed with just roses. They are mated with pigs, tomatoes with oak trees, fish with asses, butterflies with worms, orchids with snakes. The technology that makes this all possible is called biolistics-a gunshot-like violence that pierces the nuclear membrane of cells. This essentially violates not just the core chambers of life (physically crossing nuclear membranes) but the conscious-choice principle that is part of living nature's essence. Some also compare it to the violent crossing of territorial borders of countries, subduing inhabitants against their will.

What will happen if this technology is allowed to spread? Fifty years ago few predicted that chemical pollution would cause so much vast environmental harm. Now nearly $1/3^{rd}$ of all species are threatened with extinction (and up to half of all plant species and half of all mammals). Few also knew that cancer rates would skyrocket during this same period. Nowadays approximately 41% on average of Americans can expect cancer in their lifetime.

ALARM SIGNALS

No one has a crystal ball to see future consequences of the overall GMO technology. Nevertheless, there are silent alarm signals like the early death of canaries in a mine shaft. There is, for example, growing evidence that the wholesale disappearance of bees relates directly to the appearance of ever more GM pollen. If we understand certain philosophical issues about the 17th century's worldview, the potential harm of GMOs actually can potentially far outweigh that of chemical pollution.

This is because chemistry deals mostly with things altered by fire (and then no longer alive, isolated in laboratories-and not infecting living terrains in self-reproducible ways). Thus a farmer may use a chemical for many decades, and then let the land lie fallow to convert it back to organic farming. This is because the chemicals tend to break down into natural substances over time, Genetic pollution, however, can alter the oil's life forever!

Farmers who view their land as their primary financial asset have reason to heed this warning. They need to be alarmed by evidence that genetically-modified soil bacteria contamination can arise. This is more than just possible, given the numerous (1600 or more) distinct microorganisms that can be found in a single teaspoon of soil.

If that soil contamination remains permanently, the consequences can be catastrophic. Someday the public may blacklist precisely those farms that have once planted GM crops. No one has put up any warning signs on product packaging for farmers, including those who now own 1/4 of all agricultural tracks in the US. Furthermore, the spreading potential impact on all ecosystems is profound.

"Our way of life is likely to be more fundamentally transformed in the next several decades than in the previous one thousand years...Tens of thousands of novel transgenic bacteria, viruses, plants and animals could be released into the Earth's ecosystems...Some of those releases, however, could wreak havoc with the planet's biospheres."

In short these processes involve unparalleled risks. Voices from many sides echo this view. Contradicting safety claims, no major insurance company has been willing to limit risks, or insure bio-engineered agricultural products. The reason given is the high level of unpredictable consequences. Over eight hundred scientists from 84 countries have signed The World Scientist open letter to all governments calling for a ban on the patenting of life-forms and emphasizing the very grave hazards of GMOs, genetically-modified seeds and GM foods. This was submitted to the UN, World Trade Organization and US Congress. The Union of Concerned Scientists (a 1000 plus member organization with many Nobel Laureates) has similarly expressed its scientific reservations. The prestigious medical journal, Lancet, published an article on the research of Arpad Pusztai showing potentially significant harms, and to instill debate. Britain's Medical Association (the equivalent of the AMA and with over a 100,000 physicians) called for an outright banning of genetically-modified foods and labelling the same in countries where they still exist.

In a gathering of political representatives from over 130 nations, drafting the Cartagena Protocol on Biosafety, approximately 95% insisted on new precautionary approaches. The National Academy of Science report on genetically-modified products urged greater scrutiny and assessments. Prominent FDA scientists have repeatedly expressed profound fears and reservations but their voices were muted not due to cogent scientific reasons but intense political pressure from the Bush administration in its efforts to buttress and promote the profit-potentials of a nascent biotech industry.

To counterbalance this, industry-employed scientists have signed a statement in favour of genetically-modified foods. But are any of these scientists impartial? Writes the New York Times (about a similar crisis involving genetic engineering and medical applications).

"Academic scientists who lack industry ties have become as rare as giant pandas in the wild...lawmakers, bioethics experts and federal regulators are troubled that so many researchers have a financial stake [via stock options or patent participation] ...The fear is that the lure of profit could colour scientific integrity, promoting researchers to withhold information about potentially dangerous side-effects."

Looked at from outside of commercial interests, perils of genetically modified foods and organisms are multi-dimensional. They include the creation of new "transgenic" life forms-organisms that cross unnatural gene lines (such as tomato seed genes crossed with fish genes)-and that have unpredictable behaviour or replicate themselves out of control in the wild. This can happen, without warning, *inside of our bodies* creating an unpredictable chain reaction. A four-year study at the University of Jena in Germany conducted by Hans-Hinrich Kaatz revealed that bees ingesting pollen from transgenic rapeseed had bacteria in their gut with modified genes. This is called a "horizontal gene transfer." Commonly found bacteria and microorganisms in the human gut help maintain a healthy intestinal flora. These, however, can be mutated.

Mutations may also be able to travel internally to other cells, tissue systems and organs throughout the human body. Not to be underestimated, the potential domino effect of internal and external genetic pollution can make the substance of science-fiction horror movies become terrible realities in the future. The same is true for the bacteria that maintain the health of our soil-and are vitally necessary for all forms of farming-in fact for human sustenance and survival.

Without factoring in biotechnology, milder forms of controlling nature have gravitated towards restrictive monocropping. In the past 50 years, this underlies the disappearance of approximately 95% of many native grains, beans, nuts, fruits, and vegetable varieties in the United States, India, and Argentina among other nations (and on average 75% worldwide). Genetically-modified monoculture, however, can lead to yet greater harm. Monsanto, for example, had set a goal of converting *100%* of all US soy crops to Roundup Ready strains by the year 2000. If this plan were effected, it would have threatened the biodiversity and resilience of all future soy farming practices. Monsanto laid out similar strategies for corn, cotton, wheat and rice. This represents a deepest misunderstanding of how seeds interact, adapt and change with the *living* world of nature.

One need only look at agricultural history-at the havoc created by the Irish potato blight, the Mediterranean fruit fly epidemic in California, the regional citrus canker attacks in the Southeast, and the 1970's US corn leaf blight. In the latter case, 15% of US corn production was quickly destroyed. Had weather changes not quickly ensued, most all crops would have been laid waste because a fungus attached their cytoplasm universally. The deeper reason this happened was that approximately 80% of US corn had been standardized (devitalized/mechanized) to help farmers crossbreed-and by a method akin to those used in current genetic engineering. The uniformity of plants then allowed a single fungus to spread, and within four months to destroy crops in 581 counties and 28 states in the US. According to J. Browning of Iowa State University: "Such an extensive, homogeneous acreage of plants... is like a tinder-dry prairie waiting for a spark to ignite it."

The homogeneity is unnatural, a byproduct again of deadening nature's creativity in the attempt to mechanize, to grasp absolute control, and of what ultimately yields not control but wholesale disaster. Europeans seem more sensitive than Americans to such approaches, given the analogous metaphor of German eugenics.

HISTORICAL SYNOPSIS

Overall the "biotech revolution" that is presently trying to overturn 12,000 years of traditional and sustainable agriculture was launched in the summer of 1980 in the US. This was the result of a little-known US Supreme Court decision *Diamond vs. Chakrabarty* where the highest court decided that biological life could be legally patentable.

Ananda Mohan Chakrabarty, a microbiologist and employee of General Electric (GE), developed at the time a type of bacteria that could ingest oil. GE rushed to apply for a patent in 1971. After several years of review, the US Patent and Trademark Office (PTO) turned down the request under the traditional doctrine that life forms are not patentable. Jeremy Rifkin's organization, the Peoples Business Commission, filed the only brief in support of the ruling. GE later sued and won an overturning of the PTO ruling. This gave the go ahead to further bacterial gmo research throughout the 1970's. Then in 1983 the first genetically-modified plant, an anti-biotic resistant tobacco was introduced. Field trials then began in 1985, and the EPA approved the very first release of a GMO crop in 1986. This was a herbicide-resistant tobacco. All of this went forward due to a regulatory green light as in 1985 the PTO also decided the Chakrabarty ruling could be further extended to all plants and seeds, or the entire plant kingdom.

It then took another decade before the first genetically-altered crop was commercially introduced. This was the famous delayed-ripening "Flavr-savr" tomato approved by the FDA on May 18, 1994. The tomato was fed in laboratory trials to mice who, normally relishing tomatoes, refused to eat these lab-creations and had to be force-fed by tubes. Several developed stomach lesions and seven of the forty mice died within two weeks. Without further safety testing the tomato was FDA approved for commercialization. Fortunately, it ended up as a production and commercial failure, and was ultimately abandoned in 1996. This was the same year Calgene, the producer, began to be bought out by Monsanto. During this period also, and scouring the world for valuable genetic materials, W.R. Grace applied for and was granted fifty US patents on the neem tree in India. It even patented the indigenous knowledge of how to medicinally use the tree (what has since been called biopiracy). Also by the close of the 20th century, about a dozen of the major US crops-including corn, soy, potato, beets, papaya, squash, tomato and cotton-were approved for genetic modification.

Going a step further, on April 12, 1988, PTO issued its first patent on animal life forms (known as oncomice) to Harvard Professor Philip Leder and Timothy A. Stewart. This involved the creation of a transgenic mouse containing chicken and human genes. Since 1991 the PTO has controversially granted other patent rights involving human stem cells, and later human genes. A United States company, Biocyte was awarded a European patent on all umbilical cord cells from fetuses and newborn babies. The patent extended exclusive rights to use the cells without the permission of the donors. Finally the European Patent Office (EPO) received applications from Baylor University for the patenting of women who had been genetically altered to produce proteins in their mammary glands. Baylor essentially sought monopoly rights over the use of human mammary glands to manufacture pharmaceuticals. Other attempts have been made to patent cells of indigenous peoples in Panama, the Solomon Islands, and Papua New Guinea, among others.

Thus the groundbreaking Chakrabarty ruling evolved, and within little more than two decades from the patenting of tiny, almost invisible microbes, to allow the genetic modification of virtually all terrains of life on Earth.

Certain biotech companies then quickly, again with lightening speed, moved to utilize such patenting for the control of first and primarily seed stock, including buying up small seed companies and destroying their non-patented seeds. In the past few years, this has led to a near monopoly control of certain genetically modified commodities, especially soy, corn, and cotton (the latter used in processed foods when making cottonseed oil). As a result, between 70-75% of processed grocery products, as estimated by the Grocery Manufacturers of America, soon showed genetically-modified ingredients. Yet again without labelling, few consumers in the US were aware that any of this was pervasively occurring. Industry marketers found out that the more the public knew, the less they wanted to purchase GM foods. Thus a concerted effort was organized to convince regulators (or bribe them with revolving-door employment arrangements) not to require such labelling.

THE ECOLOGICAL IMPACTS OF AGRICULTURAL BIOTECHNOLOGY

Transnational corporations (TNCs) such as Monsanto, DuPont, and Novartis, the main proponents of biotechnology, argue that carefully planned introduction of these crops should reduce or even eliminate the enormous crop losses due to weeds, insect pests, and pathogens. In fact, they argue that the use of such crops will have added beneficial effects on the environment by significantly reducing the use of agrochemicals.13 However, ecological theory predicts that as long as transgenic crops follow closely the pesticide paradigm prevalent in modern agriculture, such biotechnological products will do nothing but reinforce the pesticide treadmill in agroecosystems, thus legitimizing the concerns that many environmentalists and some scientists have expressed regarding the possible environmental risks of genetically engineered organisms. In fact, there are several widely accepted environmental drawbacks associated with the rapid deployment and widespread commercialization of such crops in large monocultures.

Biotechnology may someday be considered a safe agricultural tool but studies suggest it may have harmful ecological consequences, such as:

- spreading genetically-engineered genes to indigenous plants
- increasing toxicity, which may move through the food chain
- disrupting nature's system of pest control
- creating new weeds or virus strains.

The cassava makes up part of the diet of nearly 600 million people worldwide. By inserting a bacterial version of the gene for starch production, scientists have come up with a super-sized cassava. Photo: David Monniaux.

Transgenic crops (GMCs: genetically modified crops), main products of agricultural biotechnology, are increasingly becoming a dominant feature of the agricultural landscapes of the USA and other countries such as China, Argentina, Mexico and Canada.

Nearly half of American farms grow GMCs (genetically modified crops).

- Worldwide, the areas planted to transgenic crops jumped more than twenty-fold in the past six years, from 3 million hectares in 1996 to nearly 44.2 million hectares in 2000.
- In the USA, Argentina and Canada, over half of the average for major crops such as soybean, corn and canola are planted in transgenic varieties.
- Herbicide resistant crops (HRCs) and insect resistant crops (Bt crops) accounted respectively for 59 and 15 percent of the total global area of all transgenic crops in 2000.

Big business claims GMCs will reduce the use of chemical pesticides.

Transnational corporations (TNCs) such as Monsanto, DuPont, and Novartis, the main proponents of biotechnology, argue that carefully planned introduction of these crops should reduce or even eliminate the enormous crop losses due to weeds, insect pests, and pathogens. In fact, they argue that the use of such crops will have added beneficial effects on the environment by significantly reducing the use of agro chemicals. However, ecological theory predicts that as long as transgenic crops follow closely the pesticide paradigm prevalent in modern agriculture, such biotechnological products will do nothing but reinforce the pesticide treadmill in agro ecosystems, thus legitimizing the concerns that many environmentalists and some scientists have expressed regarding the possible environmental risks of genetically engineered organisms. In fact, there are several widely accepted environmental drawbacks associated with the rapid deployment and widespread commercialization of such crops in large monocultures, including:

Pests show rapid evolution in resisting the pesticide properties of GMCs.

Toxic buildup in GMCs harms useful insects.

- the spread of transgenes to related weeds or conspecifics via crop-weed hybridization

- reduction of the fitness of non-target organisms through the acquisition of transgenic traits via hybridization
- the rapid evolution of resistance of insect pests such as Lepidoptera to Bt
- accumulation of the insecticidal Bt toxin, which remains active in the soil after the crop is ploughed under and binds tightly to clays and humic acids;
- disruption of natural control of insect pests through intertrophic-level effects of the Bt toxin on predators
- unanticipated effects on non-target herbivorous insects (*i.e.*, monarch butterflies) through deposition of transgenic pollen on foliage of surrounding wild vegetation
- vector-mediated horizontal gene transfer and recombination to create new pathogenic organisms.

This paper will focus on the known effects of the two dominant types of GMCs: herbicide resistant crops (HRCs) and insect resistant crops (Bt).

BIOTECHNOLOGY, AGRODIVERSITY AND FARMERS' OPTIONS

Monoculture, or farming only one crop, can lead to economic disaster and hunger.

The spread of transgenic crops threatens crop diversity by promoting monocultures which leads to environmental simplification and genetic erosion. History has repeatedly shown that uniformity characterizing agricultural areas sown to a smaller number of varieties is a source of increased risk for farmers, as the genetically homogeneous fields may be more vulnerable to disease and pest attack. Farmers have many choices, other than biotechnology, that work.

Several people think that HRCs and Bt crops have been a poor choice of traits to feature the technology, given predicted environmental problems and the issue of resistance evolution. In fact, there is enough evidence to suggest that both these types of crops are not really needed to address the problems they were designed to solve. On the contrary, they tend to reduce the pest management options available to farmers. There are many alternative approaches, (*e.g.*, rotations, polycultures, cover crops, biological control, etc.) that farmers can use to effectively regulate the insect and weed populations that are being targeted by the biotechnology industry. To the extent that transgenic crops further entrench the current monocultural system, they impede farmers from using a plethora of alternative methods.

ECOLOGICAL EFFECTS OF HRCS

GENE FLOW

Are we altering the genetic structure of living things in the name of utility and profit?

Just as it occurs between traditionally improved crops and wild relatives, pollen mediates gene flow between GMCs and wild relatives or conspecifics despite all possible efforts to reduce it. Little is known about the long-term persistence of crop genes in wild populations or about the impact of fitness-related crop genes on the population dynamics of weedy relatives. The main concern with transgenes that confer significant biological advantages is that they may transform wild/weed plants into new or worse weeds.

Hybridization of HRCs with populations of free living relatives will make these plants increasingly difficult to control, especially if they are already recognized as agricultural weeds and if they acquire resistance to widely used herbicides. For example:

- Transgenic resistance to glufosinate can be passed on from Brassica napus to populations of weedy Brassica napa, and persist under natural conditions.
- In Europe there is a major concern about the possibility of pollen transfer of herbicide tolerant genes from *Brassica* oilseeds to *Brassica nigra* and *Sinapis arvensis*.

ECONOMIC AND AGRONOMIC IMPLICATIONS

The majority of GMCs have been engineered to repel unwanted plants or weeds.

Worldwide in 2000, transgenic herbicide resistant crops were planted on 74% of the 44.2 million hectares devoted to transgenic crops.

In North America, transgenic glufosinate resistant cultivars of canola and corn, and transgenic glyphosate resistant cultivars of soybean, corn, cotton, and canola are now commercially available. Bromoxynil resistant transgenic cotton has also been developed. The so-called Round-up ready soybeans are the most prevalent GMCs.

Transgenic herbicide resistance in crop plants simplifies chemically based weed management because it typically involves compounds that are active on a very broad spectrum of weed species. Post-emergence application timing for these materials fits well with reduced or zero-tillage production methods, which can conserve soil and reduce fuel and tillage costs.

However, HRCs also have significant problems.

Editor's note: Sales of biotechnology products are reaching $60 billion/year.

- Reliance on HRCs perpetuates the weed resistance problems and species shifts that are common to conventional herbicide based approaches.
- Herbicide resistance becomes more of a problem as the number of herbicide modes of action to which weeds are exposed becomes fewer and fewer, a trend that HRCs may exacerbate due to market forces.

- Given industry pressures to increase herbicide sales, acreage treated with broad-spectrum herbicides will expand, exacerbating the resistance problem. For example, it has been projected that the acreage treated with glyphosate will increase to nearly 150 million acres. Although glyphosate is considered less prone to weed resistance, the increased use of the herbicide will result in weed resistance, even if more slowly, as it has been already documented with Australian populations of annual ryegrass, quackgrass, birdsfoot trefoil and *Cirsium arvense*.
- Perhaps the greatest problem of using HRCs to solve weed problems is that they steer efforts away from crop diversification and help to maintain cropping systems dominated by one or two annual species. Crop diversification can
  - reduce the need for herbicides
  - improve soil and water quality
  - minimize requirements for synthetic nitrogen fertilizer
  - regulate insect pest and pathogen populations
  - increase crop yields and reduce yield variance.

Thus, to the extent that transgenic HRCs inhibit the adoption of diversified cropping systems that include rotational crops, cover crops and green manure, they hinder the development of sustainable agriculture.

ECOLOGICAL RISKS OF BT CROPS

The number of crops engineered for insect resistance is on the rise.

Based on the fact that more than 500 species of pests have already evolved resistance to conventional insecticides, pests can also evolve resistance to Bt toxins present in transgenic crops. No one questions if Bt resistance will develop, the question is now how fast it will develop. Susceptibility to Bt toxins can therefore be viewed as a natural resource that could be quickly depleted by inappropriate use of Bt crops. However, cautiously restricted use of these crops should substantially delay the evolution of resistance. But is cautious use of Bt crops possible given commercial pressures that have resulted in a rapid rollout of Bt crops reaching 8.2 million hectares worldwide in 2000?

Organic farmers can produce a significant yield without insecticides.

The refuge strategy of setting aside 20-30% of land to non-Bt crops to delay resistance is very difficult to implement regionally. Data from the Midwest shows that Bt corn saves on some insecticide use and yields are 2.4 Bu/acre higher than conventional corn but only under high European corn borer infestations (USDA 1999). On the other hand organic corn growers use no insecticides and obtain yields (4.8-9 t/ha) similar or slightly higher than conventional farmers (5.0-7.1 t/ha). GMCs may have unintended victims, such as the monarch butterfly or lacewing.

EFFECTS ON THE SOIL ECOSYSTEM

The possibilities for soil biota to be exposed to transgenic products are very high. The little research conducted in this area has already demonstrated:

Toxins from GMCs remain active in the soil, decreasing soil fertility.

- There is long term persistence of insecticidal products (Bt and proteinase inhibitors) in soil.
- The insecticidal toxin produced by Bacillus thuringiensis subsp. kurskatki remain active in the soil, where it binds rapidly and tightly to clays and humic acids.
- The bound toxin retains its insecticidal properties and is protected against microbial degradation by being bound to soil particles, persisting in various soils for at least 234 days.
- The presence of the toxin in exudates from Bt corn and verified that it was active in an insecticidal bioassay using larvae of the tobacco hornworm.

Given the persistence and the possible presence of exudates, there is potential for prolonged exposure of the microbial and invertebrate community to such toxins, and therefore studies should evaluate the effects of transgenic plants on both microbial and invertebrate communities and the ecological processes they mediate.

If transgenic crops substantially alter soil biota and affect processes such as soil organic matter decomposition and mineralization, this would be of serious concern to organic farmers and most poor farmers in the developing world. These farmers cannot purchase or don't want to use expensive chemical fertilizers. They rely instead on local residues, organic matter and especially soil organisms for soil fertility (*e.g.*, key invertebrate, fungal or bacterial species) which can be affected by the soil bound toxin. Soil fertility could be dramatically reduced if crop leachates inhibit the activity of the soil biota and slow down natural rates of decomposition and nutrient release.

GENERAL CONCLUSIONS AND RECOMMENDATIONS

The available independently generated scientific information suggests that

The long-term impacts of GMCs are not yet known.

- the massive use of transgenic crops poses substantial potential risks from an ecological point of view *the ecological effects are not limited to pest resistance and creation of new weeds or virus strains
- transgenic crops can produce environmental toxins that move through the food chain and also may end up in the soil and water affecting invertebrates and probably ecological processes such as nutrient cycling
- no one can really predict the long-term impacts that will result from such massive deployment of such crops.

Not enough research has been done to evaluate the environmental and health risks of transgenic crops, an unfortunate trend. Most scientists feel that such knowledge is crucial to have before biotechnological innovations are implemented. There is a clear need to further assess the severity, magnitude and scope of risks associated with the massive field deployment of transgenic crops. Much of the evaluation of risks must move beyond comparing GMC fields and conventionally managed systems to include alternative cropping systems featuring crop diversity and low-external input approaches. This will allow real risk/benefit analysis of transgenic crops in relation to known and effective alternatives.

The loss of agricultural diversity may lead to disaster in developing countries.

Moreover, the large-scale landscape homogenization with transgenic crops will exacerbate the ecological problems already associated with monoculture agriculture. Unquestioned expansion of this technology into developing countries may not be wise or desirable. There is strength in the agricultural diversity of many of these countries, and it should not be inhibited or reduced by extensive monoculture, especially when consequences of doing so results in serious social and environmental problems. The repeated use of transgenic crops in an area may result in cumulative effects such as those resulting from the buildup of toxins in soils. For this reason, risk assessment studies not only have to be of an ecological nature in order to capture effects on ecosystem processes, but also of sufficient duration so that probable accumulative effects can be detected. The application of multiple diagnostic methods will provide the most sensitive and comprehensive assessment of the potential ecological impact of transgenic crops.

AGRICULTURAL BIOTECHNOLOGIES IN DEVELOPING COUNTRIES

Here we provide a brief overview of the main kinds of agricultural biotechnologies that have been used in developing countries over the past 20 years and that should be covered in the e-mail conference. They are described separately, although in practice more than one may be used together in certain situations. Note, new biotechnologies that are still at the research level, be it in the laboratory or at the field trial stage, but which have not yet been applied (*i.e.* used for commercial production by farmers) in developing countries are not included.

A short description of the different biotechnologies is provided below, indicating also what they are used for, the food and agricultural sectors involved and giving some examples of their applications in specific developing countries. Regarding the examples, their inclusion in the document does not imply that these applications have been a partial or complete success (or, conversely, that they have been any kind of a failure). Indeed, these are the kind of issues to be addressed by participants during this e-mail conference. Also, it should be kept in mind that, although not the subject of this e-mail conference, the pathway from a research development in the laboratory to its eventual application in the field (*e.g.* farmers cultivating a new genetically improved plant variety or using a new vaccine against an animal disease) can be long, resource-demanding and unsuccessful, so many biotechnologies of seemingly high promise at the experimental stage have had limited applications in developing countries so far.

As many of the biotechnologies are related to molecular biology and genetic material, some basic terminology is introduced here. Living things are made up of cells that are programmed by genetic material called DNA. A DNA molecule is made up of a long chain of nitrogen-containing bases. Only a small fraction of this DNA sequence typically makes up genes *i.e.* that code for proteins, which are molecules essential for the functioning of living cells, made up of chains of amino acids. The remaining and major share of the DNA represents non-coding sequences whose role is not yet clearly understood.

The genetic material is organized into sets of chromosomes (*e.g.* 5 pairs in Arabidopsis thaliana-a model plant species; 30 pairs in cattle), and the entire set is called the genome. In a diploid individual (*i.e.* where chromosomes are organized in pairs), there are two alleles of every gene-one from each parent-transmitted by gametes (reproductive cells) that are normally haploid (having just one of each of the pairs of chromosomes). A typical genome contains several thousand genes *e.g.* about 30,000 genes in grasses like rice and sorghum.

MOLECULAR MARKERS

Molecular markers are identifiable DNA sequences, found at specific locations of the genome, transmitted by standard Mendelian laws of inheritance from one generation to the next. They rely on a DNA assay and a range of different kinds of molecular marker systems exist, such as restriction fragment length polymorphisms (RFLPs), random amplified polymorphic DNAs (RAPDs), amplified fragment length polymorphisms (AFLPs) and microsatellites. The technology has improved in the past decade and faster, cheaper systems like single nucleotide polymorphisms (SNPs) are increasingly being used. The different marker systems may vary in aspects such as their technical requirements, the amount of time, money and labour needed and the number of genetic markers that can be detected throughout the genome.

Molecular markers have been used in laboratories since the late 1970s and they are applied across all the food and agricultural sectors. They are very versatile and can be used for a variety of purposes.

Thus, they are used in genetic improvement, through so-called marker-assisted selection (MAS), where markers physically located beside (or, even, within) genes of interest (such as those affecting yield in maize) are used to select favourable variants of the genes. MAS is made possible by the development of molecular marker maps, where many markers of known location are interspersed at relatively short intervals throughout the genome, and the subsequent testing for statistical associations between marker variants and the traits of interest. Marker maps are now available for a wide range of economically important agricultural species. Progress in the field of genomics (the study of an organism's entire genome) has also provided much useful information for MAS, enabling in some cases markers to be used that are located within the genes of interest.

Molecular markers are also used to characterize and conserve genetic resources, where some of the approaches can be applied in each of the crop, forestry, livestock and fishery sectors (*e.g.* estimating the genetic relationships between populations within a species). Other uses again are more sector-specific, such as their utilization to identify duplicate accessions in crop genebanks; monitor effective population sizes (Ne) in capture fish populations or carry out biological studies (*e.g.* of mating systems, pollen movement and seed dispersal) in forest tree populations. They are also used in disease diagnosis, to characterize and detect pathogens in livestock, crops, forest trees, fish and food.

Molecular markers have been used in a number of developing countries. In livestock, for example, they have been used in four African countries for characterization of genetic resources and in eight Asian countries, where six used them for genetic distance studies and two for MAS. In Latin America and the Caribbean, most countries have used molecular techniques, primarily for characterization purposes, while their use has been limited in the Near and Middle East. In crops, several examples of new hybrids and varieties developed through MAS are available, and in progress, in different crops, such as pearl millet, rice and maize, and in several developing countries like Bangladesh, India and Thailand. Different centres of the Consultative Group on International Agricultural Research (CGIAR) have been working with partners in developing countries to accelerate plant breeding practices through MAS.

GENETIC MODIFICATION

A genetically modified organism (GMO) is an organism in which one or more genes (called transgenes) have been introduced into its genetic material from another organism. The genes may be from a different kingdom (*e.g.* a bacterial gene introduced into plant genetic material), a different species within the same kingdom or even from the same species. For example, so-called 'Bt crops' are crops containing genes derived from the soil bacterium Bacillus thuriengensis coding for proteins that are toxic to insect pests that feed on the crops. The issue of GMOs has been highly controversial over the past decade. Many countries have introduced specific frameworks to regulate their release and commercialization.

GM crops were first grown commercially in the mid 1990s. While the majority continues to be grown in developed countries, an increasing number of developing countries are reported to be cultivating them.

Recent estimates indicate that 10 developing countries planted over 50,000 hectares of GM crops in 2008 *i.e.* Argentina (21.0 million hectares), Brazil (15.8), India (7.6), China (3.8), Paraguay (2.7), South Africa (1.8), Uruguay (0.7), Bolivia (0.6), Philippines (0.4) and Mexico (0.1). For comparison, in 1997 the only developing countries reported were Argentina (1.4 million hectares), China (1.8) and Mexico (less than 0.1). Almost all GM crops grown commercially are genetically modified for one or both of two main traits: herbicide tolerance (63% of GM crops planted in 2008) or insect resistance (15%), *i.e.* Bt crops, while 22% have both traits. Commercial release of GM forest trees has been reported in one country, China. In 2002, approval was granted for the environmental release of two kinds of Bt trees, the European black poplar (Populus nigra) and the hybrid white poplar clone GM 741, together representing about 1.4 million plants on 300-500 hectares (FAO, 2004). Regarding GM livestock or fish, there has been no commercial release for food and agriculture purposes in any developing or developed country.

Although documentation is generally quite poor, use of genetically modified microorganisms (GMMs) in the agro-industry and animal feed sector is routine in developed countries and is also a reality in many developing countries. In the agro-industry sector, use of enzymes, *i.e.* proteins that catalyse specific chemical reactions, is important. Many of the enzymes used in the food industry are commonly produced using GMMs. For example, since the early 1990s, preparations containing chymosin (an enzyme used to curdle milk in the preliminary steps of cheese manufacture) derived from GM bacteria have been available commercially. Similarly, many colours, vitamins and essential amino acids used in the food industry are also from GMMs.

In animal nutrition, feed additives such as amino acids and enzymes are widely used in developing countries. The greatest use is in pig and poultry production where, over the last decade, intensification has increased, further accelerating the demand for feed additives. For example, most grain-based livestock feeds are deficient in essential amino acids such as lysine, methionine and tryptophan and for high producing monogastric animals (pigs and poultry) these amino acids are added to diets to increase productivity. The amino acids in feed, L-lysine, DL-methionine, L-threonine and L-tryptophan, constitute over half of the total amino acid market. In India alone, the amino acid market amounts to about 5 million US dollars. The essential amino acids are produced in some cases by GM strains of Escherichia coli.

In the dairy industry, recombinant bovine somatotropin (rBST), a protein hormone from an Escherichia coli K-12 bacterium containing the cow somatotropin gene, has been used to increase milk production in a number of developing countries. Chauvet and Ochoa (1996) report that rBST was first used in Mexico in 1990 and has been sold in a number of other developing countries, including Brazil, Malaysia, South Africa and Zimbabwe.

CHROMOSOME NUMBER MANIPULATION

As mentioned earlier, genetic material is organized into sets of chromosomes and each plant and animal species has a characteristic number of chromosomes. Manipulation of whole sets of chromosomes is possible and is used for a range of different purposes in agriculture. For example, fish and shellfish have been extensively studied for manipulation of their chromosomes during early stages of development. Using relatively simple techniques such as cold and hot shocks it is possible to induce triploidy (*i.e.* with three sets of chromosomes), leading to the production of nearly completely sterile populations. Sterility may be desirable in conservation programs, where it can prevent introgression of escaped individuals from commercial stocks into natural populations. It may also be of interest in commercial fish operations, *e.g.* when developing hybrid stocks or to prevent the side effects of sexual maturation on carcass quality. As in fish, induction of sterility in crops may be desirable in certain breeding programmes, *e.g.* to produce seedless fruits, and one of the most rapid and cost-effective approaches is to create polyploids (*i.e.* with more than 2 complete sets of chromosomes), especially triploids. Triploid varieties have been produced in numerous fruit crops including most of the citrus fruits, acacias and the kiwifruit.

Another example of chromosomal number modification in fish is the production of haploid individuals after eggs are fertilized by sperm which do not contribute genetic material (a process called gynogenesis) or when normal sperm fertilize eggs whose DNA has been deactivated (a process called androgenesis). In both cases the haploid chromosomes can then be duplicated using shocks. The importance of gynogenesis/androgenesis is that it is possible to develop inbred individuals, which may be useful in fish breeding experiments aimed at producing clonal lines for detecting genomic regions affecting quantitative traits.

In crops, chromosome doubling is one of the most important technologies for the creation of fertile inter-specific hybrids (wide-crosses). Wide crossing involves hybridizing a crop variety with a distantly related plant from outside its normal sexually compatible gene pool. Its usual purpose is to obtain a plant that is virtually identical to the original crop, except for a few genes contributed by the distant relative. The technique has enabled breeders to access genetic variation beyond the normal reproductive barriers of their crops. For example, the New Rice for Africa (NERICA) hybrids are derived from crossing two species of cultivated rice, the African rice and the Asian rice, combining the high yields from the Asian rice with the ability of the African rice to thrive in harsh environments. Wide-hybrid plants are often sterile so their seed cannot be propagated, due to differences between the sets of chromosomes inherited from genetically divergent parental species that prevent stable chromosome pairing during meiosis. However, if chromosome number is artificially doubled, the hybrid may be able to produce functional pollen and eggs and be fertile. Colchicine has been used for chromosome doubling in plants since the 1940s and has been applied to more than 50 plant species, including most important annual crops. More recently, several additional chromosome doubling agents, all of which act as inhibitors of mitotic cell division, have been used in plant breeding programmes. To date, with the help of chromosome doubling technology, hundreds of new varieties have been produced worldwide.

In crops and forest trees, chromosome doubling has also been used, as for fish, to generate 'doubled haploids'. The haploid plants can be produced using anther culture which involves the in vitro culture of immature anthers. As the pollen grains are haploid, the resulting pollen-derived plants are also haploid. Doubled haploid plants were first produced in the 1960s using colchicine and today, thermal shock or mannitol incubation can be used. They may also be produced from ovule culture. Breeders value doubled haploid plants because they are 100% homozygous so any recessive genes are readily apparent. The time required after a conventional hybridization to select pure lines carrying the required recombination of characters is thus drastically reduced. Since the 1970s, doubled haploid methods have been used to create new varieties of barley, wheat, rice, melon, pepper, tobacco, and several Brassicas. In the developing world, a major centre of such breeding work is China, where numerous haploid crops have been released and many more are being developed. By 2003, China was cultivating over 2 million hectares of doubled haploid varieties, the most important of which are rice, wheat, tobacco and peppers.

BIOTECHNOLOGY-BASED DIAGNOSTICS

Applications of biotechnology for diagnostic purposes are important in crops, forest trees, livestock and fish as well as for food safety purposes. Two main kinds of methods are used, those based on the enzyme-linked immunosorbent assay (ELISA) and those based on the polymerase chain reaction (PCR).

Elisa systems are antibody-based techniques for the determination of the presence and quantity of specific molecules in a mixed sample. They are used in a range of formats, both for the detection of pathogens and for detection of antibodies produced by the host as a response to the pathogens, and a range of commercial kits are available, *e.g.* to detect fish and shrimp pathogens. Some of the ELISA-based methods use monoclonal antibodies, produced by a cell line that is both immortal and able to produce highly specific antibodies, or polyclonal antibodies, produced by many cell lines. In livestock, ELISAs form the large majority of prescribed tests for the OIE notifiable animal diseases, and many diagnostic kits are available in developing countries.

The PCR-based methods rely on the fact that each species of pathogen carries a unique DNA or RNA sequence that can be used to identify it. PCR allows production of a large quantity of a desired DNA from a complex mixture of heterogeneous sequences. It can amplify a selected region of 50 to several thousand DNA base pairs into billions of copies. After amplification, the target DNA can be identified using techniques such as gel electrophoresis or hybridization with a labelled nucleic acid (a probe). Real time PCR (or quantitative PCR) enables quantification of DNA or RNA present in a sample. The genomes of some viruses, such as the influenza A virus, are made of RNA instead of DNA, and to identify RNA from these viruses a complementary DNA (cDNA) copy of the RNA is first synthesized using an enzyme called reverse transcriptase. The cDNA then acts as the template to be amplified by PCR. This method is called reverse transcriptase PCR (RT-PCR).

PCR-based techniques offer high sensitivity and specificity and diagnostic kits allow the rapid screening of the virus or bacteria and have a direct use in situations where individuals show no antibody response after infection. For example, molluscs do not produce antibodies, and therefore antibody-based diagnostic tests are limited in their application to pathogen detection in these species. In fisheries, PCR-related tools are increasingly being used in developing countries, although they require detailed knowledge of the genomics of the pathogen itself and extensive validation in practice.

In livestock, public sector production of diagnostic kits for animal diseases in Asia and Latin America can be found in Brazil, Chile, China, India, Mexico and Thailand. Research capabilities for development, standardization and validation of diagnostic methods are also well advanced in these countries. PCR-based diagnostics are increasingly being employed in developing countries to back up findings from serological analyses. However, their use is largely restricted to laboratories of research institutions and universities and to the central and regional diagnostic laboratories run by governments. In aquaculture, there are some highly integrated companies operating in developing countries (*e.g.* in shrimp production) and these companies commonly use PCR-based diagnostic systems, where the analyses are either carried out by laboratories of the companies themselves or are outsourced to specialized private laboratories.

Biotechnology-based diagnostics are also important in food analysis. Many of the classical food microbiological methods used in the past were culture-based, with microorganisms grown on agar plates and detected through biochemical identification. These methods are often tedious, labour-intensive and slow. Genetic based diagnostic and identification systems can greatly enhance the specificity, sensitivity and speed of microbial testing. Molecular typing methodologies, commonly involving PCR, ribotyping (a method to determine homologies and differences between bacteria at the species or subspecies/strain level, using RFLP analysis of ribosomal RNA genes) and pulsed-field gel electrophoresis (a method of separating large DNA molecules on agarose gels), can be used to characterize and monitor the presence of spoilage flora (microbes causing food to become unfit for eating), normal flora and microflora in foods. RAPD or AFLP molecular marker systems can also be used for the comparison of genetic differences among species, subspecies and strains, depending on the reaction conditions used. The use of combinations of these technologies and other genetic tests allows the characterization and identification of organisms at the genus, species, subspecies and even strain levels, thereby making it possible to pinpoint sources of food contamination, to trace microorganisms throughout the food chain or to identify the causal agents of food-borne illnesses.

DEVELOPMENT OF VACCINES USING BIOTECHNOLOGIES

Immunization can be one of the most effective means of preventing and hence managing animal diseases. In general, vaccines offer considerable benefits for comparative low cost, a primary consideration for developing countries. In addition, development of good vaccines for important infectious diseases can lead to reduced use of antibiotics, which is an important issue in developing countries.

As described by Kurath (2008), biotechnology has been used extensively in the development of vaccines for aquaculture, and is applied at each of the three main stages of vaccine development *i.e.* a) identification of potential antigen candidates that might be effective in vaccines (where an antigen is a molecule, usually a protein foreign to the fish, which elicits an immune response on first exposure to the immune system by stimulating the production of antibodies specific to its various antigenic determinants. During subsequent exposures, the antigen is bound and inactivated by these antibodies) b) construction of a new candidate vaccine (where biotechnology tools can be used to produce different kinds of vaccines such as DNA vaccines, recombinant vaccines or modified live recombinant viruses. For example, a DNA vaccine is a circular DNA plasmid containing a gene for a protective antigenic protein from a pathogen of interest and c assessment of candidate vaccine efficacy, its mode of action and the host response (where *e.g.* quantitative RT-PCR can be used to examine the expression of fish genes related to immune responses).

Of the countries that responded to a recent OIE survey, 4 out of 23 and 7 out of 14 African and Asian countries respectively indicated that they produce or use animal vaccines derived from biotechnology, including experimental use as well as commercial release.

REPRODUCTIVE BIOTECHNOLOGIES

A number of reproductive biotechnologies have been applied in developing countries to influence the number (and sex) of offspring from given individuals in fish and livestock populations.

ARTIFICIAL INSEMINATION

In artificial insemination (AI), semen is collected from donor male animals, diluted in suitable diluents and manually inseminated into the female reproductive tract during oestrus (heat), to achieve pregnancy. The semen can be fresh or preserved in liquid nitrogen and then thawed. Efficiency of AI can be increased by monitoring progesterone levels, *e.g.* using ELISA, to identify non-pregnant females, and/or by oestrus synchronization, where females are treated with hormones to being them into oestrus at a desired time.

AI is widely used in developing countries. For example, in India 34 million inseminations were carried out in 2007 while about 8 million were carried out in Brazil For Africa, Asia and Latin America and the Caribbean regions, AI is mostly used for cattle production (dairy). Other species for which AI is used in all three continents are sheep, goats, horses and pigs. In addition, in Asia, AI is used for chickens, camels, buffaloes and ducks, and in Latin America and Caribbean regions for rabbits, buffaloes, donkeys, alpacas and turkeys. For the most part, semen from exotic breeds is used in local livestock populations. To a lesser extent, semen from local breeds is also used for this purpose. Most of the AI services are provided by the public sector but the contribution of the private sector, breeding organizations and NGOs is also substantial. In Africa and Asia, AI use is concentrated in peri-urban areas. Progesterone monitoring and oestrous synchronization have been applied in a number of developing countries. Applications of oestrous synchronization have been limited to some intensively managed farms where AI is routinely used.

EMBRYO TRANSFER

Embryo transfer (ET) involves the transfer of an embryo from a superior donor female to a less valuable female animal. A donor is induced to superovulate (produce several ova) through hormonal treatment. The ova obtained are then fertilized within the donor, the embryos develop and are then removed and implanted in recipient females for the remainder of the gestation period. Alternatively, the embryos can be frozen for later use.

HORMONAL TREATMENT IN AQUACULTURE

In the same way as female reproduction in livestock can be controlled by hormonal treatment, it is also an important tool in aquaculture where it is applied for 2 main purposes. The first is to control reproduction of fish and shellfish, primarily to induce the final phase of ova production in order to synchronize ovulation and to enable broodstock to produce fish early in the season or when environmental conditions suppress the spawning timing of females. Implants or injection of the hormonal compound are used extensively in salmon farming.

The second purpose is to develop monosex (single sex) populations, which can be desirable in many situations. This can be, inter alia, because one sex is superior in growth or has more desirable meat quality or to prevent sexual/ territorial behaviour. For example, female sturgeon are more valuable than males because they produce caviar. Female salmon are the more valuable sex, because sexually precocious males die before they can be harvested and salmon roe has an economic value. Male tilapia are more desirable than females because they grow twice as fast. In many fish and shellfish species, sex is not permanently defined genetically and thus it can be altered in a number of ways, including treatment with sexual hormones such as testosterone or estrogen derivatives in early stages of development. To develop all-male tilapia populations, methyltestosterone can be used while monosex trout can be produced using androgens.

SPERM/EMBRYO SEXING

In livestock, to get offspring of a desired sex (*e.g.* females are preferred for dairy animals, males for beef animals), separation of X and Y sperm (*e.g.* based on staining DNA with a fluorescent dye) for AI and sexing of embryos (*e.g.* using specific DNA probes) can be used. Although these technologies are being developed and refined in a number of research institutions, they are not used at the field level in any of the developing countries, except China.

CRYOPRESERVATION

Cryopreservation, referring to the preservation of germplasm in a dormant state by storage at ultra-low temperatures, usually in liquid nitrogen (-196 C), can be used to preserve biological material (*e.g.* seeds, sperm, embryos) of crop, livestock, forest or fish populations for potential use in the future. The technology can be used for genetic improvement purposes and for management of genetic resources. In livestock, cryopreservation has been used in a number of developing countries for ex situ conservation of animal genetic resources, including Benin, Brazil, China, India and Kenya. In fish, cryopreservation of embryos is not possible but sperm cryopreservation works for many species and has been used in carp, salmon and trout breeding, especially when the aim is to "refresh" populations that have gone through a bottleneck.

Considering crops and forest trees, about 90% of the 6 million plant accessions in genebanks, mainly crops, are stored in seed genebanks. However, storage of seeds is not an option for crops or trees that do not produce seed, such as banana, or that produce recalcitrant or non-orthodox seed (*i.e.* seed that does not survive under cold storage and/or the drying conditions used in conventional ex situ conservation), such as mango, coffee, oak and several tropical forest tree species. In these situations, as well as for long-term storage of seeds from orthodox species, cryopreservation offers an alternative strategy for ex situ conservation, although its routine use is still limited. Following plant cell, tissue or organ storage at low temperatures, plants can be regenerated. For various herbaceous (*i.e.* non-woody plants), hardwood (*i.e.* broadleaf, deciduous trees) and softwood species (*i.e.* coniferous trees), cryopreservation of a wide range of tissues and organs has been achieved. There is large scale application of shoot tip cryopreservation in fruit crop germplasm collections, such as in plum and apple. Seeds of most common agricultural and horticultural species can be cryopreserved.

TISSUE CULTURE-BASED TECHNIQUES

Tissue culture refers to the in vitro culture of plant cells, tissues or organs in a nutrient medium under sterile conditions. It has been widely used for over 50 years and is now employed to improve many of the most important developing country crops. There are a number of tissue culture-based technologies and they can be employed for a range of different purposes. Some of them, used with chromosome number manipulation.

MICROPROPAGATION

Micropropagation is the laboratory practice of rapidly multiplying stock plant material to produce a large number of progeny plants, using plant tissue culture methods. For instance, shoot tips of banana or potato are excised from healthy plants and cultivated on gelatinized nutrient media in sterile conditions (in test tubes, plastic flasks, or baby food jars), so that contamination with pests and pathogens is avoided. The obtained plantlets can be multiplied an unlimited number of times, by cutting them in single-node pieces and cultivating the cuttings in similar aseptic conditions. Millions of plantlets can be produced this way in a very short time. The plantlets are then transplanted in the field or nurseries, where they grow and yield low-cost, disease-free propagation materials, ready to be distributed to farmers. Even if healthy plants are not available initially, specific in vitro techniques can also be applied to produce disease-free propagation material.

Today, micropropagation is widely used in a range of developing country subsistence crops including banana, cassava, potato and sweetpotato; commercial plantation crops, such as oil palm, coffee, cocoa, sugarcane and tea; niche crops such as cardamom and vanilla; and fruit trees such as almond, citrus, coconut, mango and pineapple. Some of the many countries with significant crop micropropagation programs include Argentina, Cuba, Gabon,

India, Indonesia, Kenya, Nigeria, Philippines, South Africa, Uganda and Vietnam.

IN VITRO SLOW GROWTH STORAGE

Micropropagation procedures have been developed for over 1,000 plant species, many of which are today micropropagated commercially. The procedures include rapid multiplication, involving rapid growth and frequent subculture (regeneration) which is generally the objective of commercial micropropagation. Instead, the basis of successful in vitro storage of stock cultures is to increase the interval between subcultures by retarding the growth without any deleterious effects on the plants in culture. The strategy is used to conserve plant genetic resources and in vitro slow growth procedures can be used so that plant material can be held 1-15 years under tissue culture conditions with periodic subculturing, depending on the species. Normally, growth is limited using low temperatures often in combination with low light intensity or even darkness. Temperatures in the range of 0-5 °C are employed for cold-tolerant species and 15-20 °C for tropical species. Growth can also be limited by modifying the culture medium and reducing oxygen levels available to the cultures.

IN VITRO EMBRYO RESCUE

Wide crossing has only become possible by advances in plant tissue culture. A particular challenge was to overcome the biological mechanisms that normally prevent inter-specific and inter-genus crosses, as a high proportion of wide-hybrid seeds either do not develop to maturity or do not contain a viable embryo. To avoid spontaneous abortion, embryos are removed from the ovule at the earliest possible stage and placed into culture in vitro. Mortality rates can be high, but enough embryos normally survive the rigours of removal, transfer, tissue culture, and regeneration to produce adult hybrid plants for testing and further crossing.

First-generation, wide-hybrid plants are rarely suitable for cultivation because they have only received half of their genes from the crop parent. From the other (non-crop) parent they have received, not only the small number of desirable genes, but also thousands of undesirable genes that must be removed by further manipulation. This is achieved by crossing the hybrid with the original crop plant, plus another round of embryo rescue, to grow up the new hybrids. This 'backcrossing' process is repeated for about six generations (sometimes more), until a plant is obtained that is almost identical to the original crop parent, except that it now contains a small number of desirable genes from the non-crop parent plant. Wide-crossing programs can take more than a decade to complete, although MAS and anther culture can be used to speed up the process. Embryo rescue has been used occasionally in forest tree species, but its application is likely to be limited to a small number of hybrids of interest, which are sufficiently close to produce a normal embryo but where embryo development in vivo is a limiting factor.

MUTAGENESIS

This involves the use of mutagenic agents, such as chemicals or radiation, to modify DNA and hence create novel phenotypes. Induced mutagenesis has been used in crop breeding programs in developing countries since the 1930s. It also includes somaclonal mutagenesis, involving changes in DNA induced during in vitro culture. Somaclonal variation is normally regarded as an undesirable by-product of the stresses imposed on a plant by subjecting it to tissue culture. However, provided they are carefully controlled, somaclonal changes in cultured plant cells can generate variation useful to crop breeders. In forestry, use of somaclonal variation has been a popular subject for research, particularly during the 1980s, but the technology is generally seen to offer little for the genetic improvement of most major industrial forest tree species.

Almost 3,000 new crop varieties have been developed and released by countries using mutation-assisted plant breeding strategies and an estimated 100 countries currently use induced mutation technology. Case studies from Kenya (wheat), Peru (barley), sub-Saharan Africa (cassava) and Vietnam (rice) are described in IAEA (2008).

In the livestock sector, mutagenesis has also been used in developing countries. The sterile insect technique (SIT) for control of insects (*e.g.* screwworm and tsetse flies) relies on the introduction of sterility in the females of the wild population. The sterility is produced following the mating of females with released males carrying, in their sperm, dominant lethal mutations that have been induced by ionizing radiation. This method is usually applied as part of an area-wide integrated pest management approach and has been applied in developing countries in the livestock sector as well as for the control of crop pests. An estimated 30 countries use SIT against insect pests, including Chile and Peru.

Mutagenesis is also extensively used to improve the quality of microorganisms and their enzymes or metabolites used in food processing. The process involves the production of mutants through the exposure of microbial strains to mutagenic chemicals or ultraviolet rays. Improved strains thus produced are selected on the basis of specific properties such as improved flavour-producing ability or resistance to bacterial viruses.

FERMENTATION

Fermentation is the process of bioconversion of organic substances by microorganisms and/or enzymes of microbial, plant or animal origin. During fermentation, various biochemical activities take place leading to the break down of complex substances into simple substances and resulting in the production of a diversity of metabolites including simpler forms of proteins, carbohydrates, fats, such as sugars, amino acids, lipids, as well as new compounds such as antimicrobial compounds (*e.g.* lysozyme, bactericins); organic acids (*e.g.* lactic acid, acetic acid, citric acid); texture-forming agents (*e.g.* xanthan gum); and flavours (esters and aldehydes). Apart from the various new products that are yielded during fermentation, the process is widely known for its preservative benefits. The new products that emerge following fermentation have been found to have potential for longer shelf lives, and they have characteristics quite different from the original substrates from which they are formed.

Fermentation is globally applied to preserve a wide range of raw agricultural materials (cereals, roots, tubers, fruit and vegetables, milk, meat and fish, etc.). Commercially produced fermented foods which are marketed globally include dairy products (cheese, yogurt, fermented milks), sausages and soy sauce. Fermentation of sugars is also central to production of bioethanol from agricultural feedstocks.

Certain microorganisms associated with fermented foods, in particular strains of the Lactobacillus species, are probiotic *i.e.* used as live microbial dietary supplements or food ingredients that have a beneficial effect on the host by influencing the composition and/or metabolic activity of the flora of the gastrointestinal tract. They can also be used as feed additives for monogastric and ruminant animals, and have been applied for this purpose in China, India and Indonesia.

In developing countries, fermented foods are produced generally at the household and village level, using traditional processes that are uncontrolled and dependent on spontaneous 'chance' microorganisms from the environment. Modern fermentation processes employ the use of well constructed vessels (fermenters/bioreactors), with appropriate controlled mechanisms for temperatures, pH, nutrients levels, oxygen tensions among others and also use selected microorganisms and/or enzymes for their operations.

BIOFERTILISERS

Soils are dynamic living systems that contain a variety of microorganisms such as bacteria, fungi and algae. Maintaining a favourable population of useful microflora is important from a fertility standpoint. The most commonly exploited microorganisms are those that help in fixing atmospheric nitrogen for plant uptake or in solubilizing/mobilizing soil nutrients such as unavailable phosphorus into plant-available forms, in addition to secreting growth-promoting substances for enhancing crop yield. As a group, such microbes are called biofertilisers or microbial inoculants. They can be generally defined as preparations containing live or latent cells of efficient strains of nitrogen-fixing, phosphate-solubilizing or cellulolytic microorganisms used for application to seed or soil with the objective of increasing the numbers of such microorganisms and accelerating certain microbial processes to augment the availability of nutrients in a form that plants can assimilate readily. Biofertilisers have been used in a number of developing countries, such as Kenya and Thailand, often involving nitrogen-fixing Rhizobia bacteria.

BIOPESTICIDES

Living organisms that are harmful to plants and cause biotic stresses are collectively called pests, and they cause tremendous economic damage to plant production worldwide. Biopesticides are mass-produced, biologically based agents used for the control of plant pests. They can be living organisms, such as microorganisms, or naturally occurring substances, such as plant extracts or insect pheromones. Microorganisms used as biopesticides include bacteria, protozoa, fungi and viruses and they are used in a range of different crops.

For example, different biopesticides are available for controlling locusts. As an illustration, a biopesticide containing spores of the fungus Metarhizium anisopliae, was used to control a migratory locust infestation in an FAO project in 2007 in Timor-Leste. Surveys revealed that an area of about 20,000 hectares was infested with gregarious nymphs and that there was a serious threat to the rice crop. The target area was considered unsuitable for chemical spraying because of high density human settlement and many water courses, so the infestation was treated with the biopesticide, targeting flying swarms using a helicopter, spraying in a time period of over one month. Note, biopesticides generally have a slow action compared to conventional chemicals and, for that reason, the latter are preferred if crops are under immediate threat.

4

Economic Benefits of Agricultural Biotechnology

INTRODUCTION

Agricultural biotechnology is any technique in which living organisms, or parts of organisms are altered to make or modify agricultural products, to improve crops, or develop microbes for specific uses in agricultural processes. Simply put, when the tools of biotechnology are applied to agriculture, it is termed as "agricultural biotechnology". Genetic engineering is also a part of agricultural biotechnology in today's world. It is now possible to carry out genetic manipulation and transformation on almost all plant species, including all the world's major crops.

Plant transformation is one of the tools involved in agricultural biotechnology, in which genes are inserted into the genetic structure or genome of plants. The two most common methods of plant transformation are Agrobacterium Transformation-methods that use the naturally occurring bacterium; and Biolistic Transformation – involving the use of mechanical means. Using any of these methods the preferred gene is inserted into a plant genome and traditional breeding method followed to transfer the new trait into different varieties of crops.

Production of food crops has become much cheaper and convenient with the introduction of agricultural biotechnology. Specific herbicide tolerant crops have been engineered which makes weed control manageable and more efficient. Pest control has also become more reliable and effective, eliminating the need for synthetic pesticides as crops resistant to certain diseases and insect pests have also been engineered. Phytoremediation is the process in which plants detoxify pollutants in the soil, or absorb and accumulate polluting substances out of the soil. Several crops have now been genetically engineered for this purpose for safe harvest and disposal, and improvement of soil quality. According to the USDA (United States Department of Agriculture)'s National

Agricultural Statistics Service (NASS), in reference to a section specific to the major biotechnology derived field crops, out of the whole crop plantings in the United States in 2004, biotechnology plantings accounted for about 46 percent for corn, 76 percent for cotton, and 85 percent for soybeans.

Modern agricultural biotechnology has now become a very well-developed science. The use of synthetic pesticides that may be harmful to man, and pollute groundwater and the environment, has been significantly lessened with the introduction of genetically engineered insect-resistant cotton. Herbicide-tolerant soybeans and corn have also enabled the use of reduced-risk herbicides that break down more quickly in soil. These are nontoxic to plants or animals, and herbicide-tolerant crops help preserve topsoil from erosion since they thrive better in no-till or reduced tillage agriculture systems. Papayas resistant to the ringspot virus were also developed through genetic engineering, which saved the U.S. papaya industry. Agricultural biotechnology may also be helpful in improving and enhancing the nutritious quality of certain crops. For example, enhancing the levels of beta-carotene in canola, soybean, and corn improves oil compositions, and reduces vitamin A deficiencies in rice. There are also researches going on in the field of biotechnology to produce crops that will not be affected by harsh climates or environments and that will require less water, fertilizer, labour etc. This would greatly reduce the demands and pressures on land and wildlife.

Where would we have been had it not been for the discovery of the art of collecting seeds and cultivating them for food? Life would surely have been very different! Early man was a food gatherer, depending on nature for all his needs. He gradually moved on to being a food grower with the discovery of agriculture, and settled down in one place, learning to live in a group. This was the beginning of civilization as we know it today.

Early knowledge of agriculture was an accumulation of experiences that were passed on from father to son. Some of these have been preserved as religious commandments and some in the ancient inscriptions. There is evidence to show that as early as 2000 BC the Egyptian civilization followed particular dates for sowing and reaping. Some Greek and Roman classics give instructions on how to get a higher yield.

The development of agriculture made it apparent that more food could be extracted from a given area of land by encouraging useful and hardy plant and animal species, and discouraging others.

At the turn of the 19th century, a movement began in central Europe to train farmers in specific farming skills. A truly scientific approach was begun by Justine von Liebig of Darmstadt who in his classic work introduced the systematic development of agriculture Science. From the 19th century onwards plant production became a scientific discipline.

In the early 20th century, the legendary work of Gregor Mendel laid the foundation of modern day genetics. His work explained the basics of inheritance in terms of the factor we today call genes. Apart from selection and hybridization, new and innovative techniques such as genetic engineering that aid plant breeders have been developed in the recent past. One example of this is BtCotton. With advances in human and plant biology, more intricate details about the cell – the basic unit of life – were illuminated. The possibility of raising whole plants from various plant tissues, commonly know as tissue culture, has thrown open the doors for expedited evolution both in terms of generation of genetic variability and multiplication of elite plant types. The knowledge of the wonder molecule DNA has also opened a new area of plant breeding research. These new technologies have been collectively referred to as biotechnology. It is a collective effort for plant breeding in the future and will compliment man's crusade for more and better food. In India, the Green Revolution saw the rapid progress of agriculture and the application of different methods to enhance production. Biofertilizers have been proven to be more environmentally friendly fertilizers that do not cause harm to life. Bioremediation methods have been used to clear oil spills using bacteria.

Biotechnology is short for biological technology. Technology is the ability to better utilize our surroundings. Biotechnology applies the same principles to living organisms as do other technologies. Biotechnology can be defined as the application of our knowledge and understanding of biology to meet practical needs. It is as old as the growing of crops. Today's biotechnology is largely identified with applications in medicine and agriculture based on our knowledge of the genetic code of life. Fermentation, used in making bread, beer, and cheese, is an example of biotechnology. Modern biotechnology simply allows scientists to be more specific in their work.

Different types of crops have been produced using the molecular tools of biotechnology and are beginning to be utilized in agricultural systems all over the world. At the same time, an increasing number of farmers are adopting sustainable cultural practices.

Biotechnology has the potential to assist farmers in reducing on-farm chemical inputs and produce value-added commodities. Conversely, there are concerns about the use of biotechnology in agricultural systems including the possibility that it may lead to greater farmer dependence on the providers of the new technology.

PLANT BREEDING

Plant breeding is the art and science of changing the genetics of plants in order to produce desired characteristics. Plant breeding can be accomplished through many different techniques ranging from simply selecting plants with desirable characteristics for propagation, to more complex molecular techniques. Plant breeding has been practiced for thousands of years, since near the beginning of human civilization. It is now practiced worldwide by individuals such as gardeners and farmers, or by professional plant breeders employed by organizations such as government institutions, universities, crop-specific industry associations or research centers. International development agencies believe that breeding new crops is important for ensuring food security by developing new varieties that are higher-yielding, resistant to pests and diseases, drought-resistant or regionally adapted to different environments and growing conditions.

BREED PLANT OR ANIMALS

As a part of agriculture, man started rearing plants and animals to meet his requirements. This is when humans started to learn how to influence the process of natural evolution so as to breed plant or animals. Slowly and gradually, this process of expedited evolution, through selection and cultivation of plants, acquired the form of a routine endeavor—what we today call 'plant breeding'. In this, heredity, which refers to the passage of various characteristic features from the main plant (the parent) to the plantlets (the progeny), plays an important role. The effects of heredity had been apparent to early man and he had taken advantage of them ever since the advent of agriculture.

Various methods have evolved in plant breeding. One of the most important methods is that of selection. The ability to choose gave birth to the idea of selection. This is the most primitive and by and large the most successful method of plant breeding. Selection as a part of plant breeding started with the domestication of plants by early man. Domestication refers to the process of bringing wild species under human management. Not all selection over the years have been human influenced—many of the important crop species have resulted from the natural selection process, which is an integral part of evolution. As human knowledge of agriculture grew, man started shuffling crop species from one geographical terrain to another, thus making new introductions.

The first prerequisite of selection is the availability of variability, *i.e.* different types of forms. After a variable population is recognized, individuals that are the best performers for the desired feature, say fruit size in the case of tomatoes, are chosen and the rest of the population is discarded or rejected. The progeny of the selected individuals is grown further and again screened for the desired feature. This process is repeated until a uniform plant population is attained which has the best-desired characters. Eventually, a desired uniform crop variety is produced by this successive selection followed by multiplication of the selected individuals.

Selecting higher yielding plant varieties is no easy task. Various tools have been devised to deal with plant selection. In fact, the birth of genetics as an independent discipline in plant Science started with some clever mathematical computations. This brainchild of yesteryears is now an important branch of genetics known as biometrics. Biometrics is defined as the application of statistics in biology.

This has contributed greatly to the development of various systems based on which selection of plants is done. There are various methods by which plant selection is carried out, namely selection for uniform plants, known as pure line selection; selection from field-grown plants, known as bulk selection or mass selection; and selection from a well-documented list of parentage, commonly known as the pedigree system. Overall, the hallmark of selection lies in human ability to chose the best plants from a cluster of many.

HYBRIDIZATION

In traditional terms, hybridization refers to the union of the male and the female gamete to produce a zygote. In plant Science, hybridization also refers to the crossing or mating of two plants. The story of scientific hybridization of crop plants started with J G Kolreuter, who in 1761 published his work on the scientific bases of hybridization. Since then, hybridization followed by selection, has been the major tool of plant breeding.

In his quest to find more variability, man started experimenting with hybridization of plants so as to achieve the perfect plant type. This process was actually the beginning of expedited evolution since it led to the formation of new plant types artificially or due to human intervention at a much faster pace than it would have happened in nature. For example, the bread wheat that we eat today has taken about 500 years to evolve to its present form through human intervention. This form of wheat would have taken thousands of years to evolve had it been left to the natural evolution process.

WAYS IN WHICH HYBRIDIZATION IS USED

Some of the ways in which hybridization has been exploited in breeding crop plants are given below

COMBINATION BREEDING

The main aim of combination breeding is to transfer one or more characters into a single variety or plant type from many others. For this, an existing plant variety may be used as the recipient parent while many other crop varieties or wild relatives may contribute as donor parents. The most commonly used method to achieve this goal is known as the backcross method. The plant type in which the character or the trait is being transferred is known as the recipient parent and the other as the donor parent. For this, the two plants are mated or crossed and the progeny is screened for the desired trait. The progeny plants possessing the desired trait are then selected and crossed back to the recipient parent. This process is repeated until the desired plant type having all the characteristics of the recipient in addition to the trait being transferred is finally obtained. This exercise is known as backcrossing. Backcrossing involves both hybridization and selection.

HYBRID VARIETIES

Plant scientists exploit the characteristic feature of better yielding 'hybrids' in plants. Hybrid vigour, or hetrosis as it is scientifically known, exploits the fact that some offspring from the progeny of a cross between two known parents would be better than the parents themselves. Many hybrid varieties of several crop species are being grown all over the world today. An example of this is the hybrid tomatoes that we eat commonly. The philosophy of hybridization has been extended from 'within the same species or genera (the same type of plants)' to 'different species or genera (totally different plants)'. This is known as wide or distant hybridization. Wide hybridization has helped breeders to break what is known as the species or genera barrier for gene transfer, *i.e.* it has helped breeders to transfer beneficial characteristics from wild and weedy plants to the cultivated crop species.

THE GREEN REVOLUTION

The world's worst recorded food disaster occurred in 1943 in British-ruled India. Known as the Bengal Famine, an estimated 4 million people died of hunger that year in eastern India (which included today's Bangladesh). Initially, this catastrophe was attributed to an acute shortfall in food production in the area. However, Indian economist Amartya Sen (recipient of the Nobel Prize for Economics, 1998) has established that while food shortage was a contributor to the problem, a more potent factor was the result of hysteria related to World War II, which made food supply a low priority for the British rulers.

When the British left India in 1947, India continued to be haunted by memories of the Bengal Famine. It was therefore natural that food security was one of the main items on free India's agenda. This awareness led, on one hand, to the Green Revolution in India and, on the other, legislative measures to ensure that businessmen would never again be able to hoard food for reasons of profit.

The Green Revolution, spreading over the period from1967/68 to 1977/78, changed India's status from a food-deficient country to one of the world's leading agricultural nations. Until 1967 the government largely concentrated on expanding the farming areas. But the population was growing at a much faster rate than food production. This called for an immediate and drastic action to increase yield. The action came in the form of the Green Revolution. The term 'Green Revolution' is a general one that is applied to successful agricultural experiments in many developing countries. India is one of the countries where it was most successful. There were three basic elements in the method of the Green Revolution:

- Continuing expansion of farming areas
- Double-cropping in the existing farmland
- Using seeds with improved genetics.

The area of land under cultivation was being increased from 1947 itself. But this was not enough to meet the rising demand. Though other methods were required, the expansion of cultivable land also had to continue. So, the Green Revolution continued with this quantitative expansion of farmlands.

Double cropping was a primary feature of the Green Revolution. Instead of one crop season per year, the decision was made to have two crop seasons per year. The one-season-per-year practice was based on the fact that there is only one rainy season annually. Water for the second phase now came from huge irrigation projects. Dams were built and other simple irrigation techniques were also adopted.

Using seeds with superior genetics was the scientific aspect of the Green Revolution. The Indian Council for Agricultural Research (which was established by the British in 1929) was reorganized in 1965 and then again in 1973. It developed new strains of high yield variety seeds, mainly wheat and rice and also millet and corn.

The Green Revolution was a technology package comprising material components of improved high yielding varieties of two staple cereals (rice and wheat), irrigation or controlled water supply and improved moisture utilization, fertilizers, and pesticides, and associated management skills.

Benefits

Thanks to the new seeds, tens of millions of extra tonnes of grain a year are being harvested.

The Green Revolution resulted in a record grain output of 131 million tonnes in 1978/79. This established India as one of the world's biggest agricultural producers. Yield per unit of farmland improved by more than 30% between1947 (when India gained political independence) and 1979. The crop area under high yielding varieties of wheat and rice grew considerably during the Green Revolution.

The Green Revolution also created plenty of jobs not only for agricultural workers but also industrial workers by the creation of related facilities such as factories and hydroelectric power stations.

Shortcomings

In spite of this, India's agricultural output sometimes falls short of demand even today. India has failed to extend the concept of high yield value seeds to all crops or all regions. In terms of crops, it remains largely confined to foodgrains only, not to all kinds of agricultural produce.

In regional terms, only the states of Punjab and Haryana showed the best results of the Green Revolution. The eastern plains of the River Ganges in

West Bengal also showed reasonably good results. But results were less impressive in other parts of India.

The Green Revolution has created some problems mainly to adverse impacts on the environment. The increasing use of agrochemical-based pest and weed control in some crops has affected the surrounding environment as well as human health. Increase in the area under irrigation has led to rise in the salinity of the land. Although high yielding varieties had their plus points, it has led to significant genetic erosion.

Since the beginning of agriculture, people have been working to improving seed quality and variety. But the term 'Green Revolution' was coined in the 1960s after improved varieties of wheat dramatically increased yields in test plots in northwest Mexico. The reason why these 'modern varieties' produced more than traditional varieties was that they were more responsive to controlled irrigation and to petrochemical fertilizers. With a big boost from the international agricultural research centres created by the Rockefeller and Ford Foundations, the 'miracle' seeds quickly spread to Asia, and soon new strains of rice and corn were developed as well.

By the 1970s the new seeds, accompanied by chemical fertilizers, pesticides, and, for the most part, irrigation, had replaced the traditional farming practices of millions of farmers in developing countries. By the 1990s, almost 75% of the area under rice cultivation in Asia was growing these new varieties. The same was true for almost half of the wheat planted in Africa and more than half of that in Latin America and Asia, and more than 50% of the world's corn as well. Overall, a very large percentage of farmers in the developing world were using Green Revolution seeds, with the greatest use found in Asia, followed by Latin America.

ECONOMIC GROWTH AND DEVELOPMENT IN AGRICULTURAL BIOTECHNOLOGY

Technological change in agriculture has been the foundation of economic growth and development. By raising yields and continually increasing the number of people who can be fed by the output of one worker in agriculture, change in agricultural technology has permitted an increasing proportion of labour and other productive resources to be devoted to non-agricultural production. Plant and animal breeding and chemical and mechanical technology plus improved husbandry have caused continual structural change in farming.

These technological stimuli have arisen as a series of overlapping waves to create a process of technological change, the momentum of which looks likely to be maintained by intensified adoption of information and biotechnologies. As a result of the cumulative processes to date, farms have become larger and more specialized, and farmers are more highly trained and have substituted machines for animal and human power; farming in the 'West' has become one of the most capital-intensive of industries.

At the same time agriculture has become more dependent upon industry, and even more industrialized-the intensive production of eggs, broiler chickens, horticultural products and pigs has many characteristics which belong to industrial production rather than to traditional farming. Goodman *et al.* talk of 'the industrial appropriation of the rural processes'. Processes of marketing farm products have passed off the farm to be performed by sophisticated distribution and retailing sectors which increasingly dictate details of production to the farm sector, while at the other end of the chain an increasing proportion of inputs is supplied by the chemical, pharmaceutical and engineering industries.

The process of 'industrializing' agriculture has been transferred to the less-developed countries (LDCs) and is epitomized by the so-called 'Green Revolution'. This has entailed expansion of irrigation systems using modern pumps, engineered dams and canals, plus the increased use of inorganic fertilizer, insecticides, herbicides, fungicides and power machinery; although at its heart has been the 'old' biotechnology of breeding new cereal varieties which give high yields when supplied with chemical inputs. Because much of the impetus for this 'revolution' has been western technology and farming know-how, particularly that of the USA, there are those who are critical of what they see as the LDCs' increasing technological dependence upon the West, its industries and research institutions.

What is undeniably the case is that technological advance has been faster, and public and private expenditure on research in western agriculture greater than in LDCs taken as a whole. Thus exportable surpluses from the West have increased with generous public support, while many LDCs have had to resort increasingly to food imports to meet their needs.

While at this stage it is difficult to foresee exactly when, and on what commercial scale, the new agricultural biotechnology 'revolution' which is now brewing will have its effect, most commentators agree that it will intensify the industrialization of agriculture, and that it will increase the technological dependence of most LDCs upon western firms. It may cause considerable disruption to the economies of some countries and will cause farming operations to diverge increasingly from the pattern associated with the country yeoman.

Rural life in many more remote farming areas will be further threatened by the continued deterioration in the economics of ruminant livestock farming on upland and low-productivity pastures. It is these redistributive and structural effects that are the central focus here rather than any attempt to estimate the scale and speed of uptake of specific biotechnology inputs and processes. Although prospective biotechnological change in agriculture is not different in many fundamental respects from other processes of technological change, in that it will operate through the provision of new or modified inputs (seeds, hormones, etc.) and the creation of new markets in industry, it does raise new issues. Are there dangers from releasing life forms which are engineered using new methods which are quite different from those associated with varieties developed by traditional plant-and animal-breeding methods? Even if the dangers can be objectively assessed and turn out to be minimal, will what may be perceived as man-made and therefore 'unnatural' products be acceptable to society as a whole?

PRINCIPAL DIRECTIONS OF BIOTECHNOLOGICAL DEVELOPMENT

In a recent review of agricultural technology, the Office of Technology Assessment (OTA) of the US Congress defined biotechnology to include 'any technique that uses living organisms or processes to make or modify products, to improve plants or animals or to develop micro-organisms for specific uses-it focuses upon recombinant DNA and cell fusion technologies'. Longworth concurs with this definition, with the significant addition of tissue-culture techniques, an aspect of biotechnology which has great economic, and therefore social and political, potential impact. Despite debates about whether these or any other definitions are adequate, they do touch on the main aspects of biotechnology which deserve to be considered.

The aspect of this new biotechnology which most captures the imagination and stirs the greatest controversy is gene splicing or recombinant DNA techniques, which inspire researchers to consider the possibilities of producing reproducible animals and plants markedly different from current existing species, and referred to as transgenic species. Already in higher animals and plants recombinant DNA has produced transgenic forms which are being commercially exploited. Their agricultural significance is, however, so far limited, and commercial application is concentrated in highly profitable pharmaceutical and horticultural markets. Genes have been introduced into several animal species which alter their protein synthesis to enable transgenic sheep to produce insulin in their milk, and rabbits to produce interferon.

A completely different application has originated in Denmark for salmon, where it has proved possible to introduce germplasm which enables the salmon's physiology to handle heavy metals, which are normally toxic, so opening up new locations for farm fisheries. Much current research is directed to conferring disease immunity on animals, and holds out the prospect of widespread commercial application. In plants one achievement has been the transfer of genetic resistance to antibiotics in the petunia, and another has been the introduction of storage-protein genes from French bean plants into tobacco plants. As yet commercial progress with recombinant-DNA technology in plants appears limited, but extensive opportunities beckon, particularly because plant research is less restricted by the ethical and animal-welfare concerns which apply to research on transgenic animals. The most important commercial developments based on gene splicing have so far occurred with genetically much simpler microorganisms, and it is with these that the greatest short-to mediumrun commercial potential lies. Already genetically engineered micro-organisms are producing a variety of hormones, vaccines, enzymes and other proteins. Important examples are the production of insulin, vaccines for neo-natal diarrhoea in calves and piglets, and the bovine growth hormone BST identical to that produced in cows which can stimulate a 20 per cent increase in milk yield. BST is already a source of problems for legislators in the European Community, and has provoked strong reactions from the media and milk consumers.

As far as larger animals, and cattle in particular, are concerned, it is developments in embryo transfer and in many processes for manipulating reproduction which hold out the prospect of continuing increases in yields, feed conversion efficiency and general economic efficiency. Already, apparently over 1 per cent of dairy calves in the USA are from embryo transplants, despite the high costs still associated with this procedure. Widespread adoption of these sophisticated technologies would further distance livestock farming from its traditional rural simplicity, and from the natural mating of animals to produce offspring. It would place technical demands upon the operators which favour large-scale company farms capable of supporting a range of highly trained specialists.

Longworth identifies new techniques of tissue and cell culture as having 'the potential for enormous advances in crop improvement in the next couple of decades'. Those techniques 'can both increase the genetic diversity and greatly increase selection efficiency', and they permit innumerable plants to be reproduced asexually from single cells or small pieces of tissue. The particular technique known as 'callus culture' has been used for many years to clone highly valued horticultural plants such as orchids, and cloning of cuttings is widely practised for tree crops such as tea and palm oil as well as by millions of gardeners for garden plants.

According to Longworth 'cell culture' of single cells has unexpectedly, and so far inexplicably, resulted in plants with different properties being regenerated from the same clump of parent tissue. Sugar cane, maize and potato plants regenerated in this way have been found which are resistant to important pathogens. This application of cell culture with the capacity to generate vast numbers of seedlings rapidly has the potential to simplify greatly the hitherto labourious procedures of plant breeding and selection.

However, in terms of current and immediate commercial importance, it is through micro-organisms that biotechnology has its greatest impact. Microbial fermentation processes have been of great commercial significance for centuries, for example in bread, wine and cheese production, and there is the prospect of considerable development. Two recent examples of new processes indicate the sorts of impact that such developments can have. In the late 1960s genetically engineered bacteria were developed which were able to digest corn starch to produce high-fructose corn syrup (HFCS) and which left as a residue corn gluten, which is now an important protein feed for livestock. HFCS has made considerable inroads in the United States and in Japan. It has largely replaced sugar as a sweetener in Coca Cola as well as many other food and drink products.

This has helped depress sugar prices to Third World growers. In the European Community steps have been taken to prevent HFCS and other new powerful sweeteners from being produced and from undermining the market for domestically grown beet sugar, which is supported by the Common Agricultural Policy. A second important application of microbial fermentation has been the production of ethanol from sugar cane in Brazil and from maize in the USA. The commercial viability of these processes is critically dependent upon the price of oil as the main non-renewable source of fuel, and, to date, massive subsidies have been required to maintain the Brazilian and USA ethanol programmes.

Longworth states, however, that a new biotechnology, Sucrotech, is being patented, which will not only reduce the cost of producing ethanol from sugar cane, but will simultaneously produce fructose at a cost which will be competitive with HFCS and may thereby reclaim part of the sweetener market for sugar cane. That such possibilities are in prospect indicates how volatile the future might be; biotechnology has tilted the competitive balance from sugar cane to maize, causing economic pressure and even disruption to sugar-cane-dependent economies and may in future switch it back again.

In the long run the capacity to ferment fuels microbially from renewable agricultural feedstocks points to an important long-term reorientation of agriculture if non-renewable oil becomes uncompetitively expensive as a fuel for cars and feedstock for certain chemicals. It suggests that eventually there will be an increased emphasis on agricultural production of industrial feedstocks at the same time as continually increasing food output will be required to feed the expanding world population. All this will require considerable increases in agricultural productivity, to which biotechnology will increasingly contribute, but it will at the same time impose great strains on the natural environment and the structure of agriculture.

PRIVATE-SECTOR CONTROL OF BIOTECHNOLOGY DEVELOPMENT

It has been the role of the public sector to undertake research and development of agricultural technology as a public service to firms which might profit from translating that R & D into a commercial product or process, to farmers who might profit from adoption of the technology, and perhaps most importantly to consumers at home and abroad who benefited from the lower prices resulting from greater abundance. This was particularly true of the phase of change dominated by improvements in plant and animal breeding, the 'old' biotechnology. While the public sector had a role in basic research for chemical and mechanical technologies, the benefits of research expenditure in these areas were easier to capture by private companies investing in R & D, so that increasingly the public sector has taken a smaller role in these areas although maintaining a strong regulatory role in regard to agricultural chemicals in particular.

Where it is impossible to prevent others from escaping payment for the research costs, either because the product is easily copied (seeds which can be regenerated by farmers or other firms) or because proposals for new methods can be readily implemented, private firms are understandably unwilling to invest. Machines, insecticides, fungicides and other manufactured inputs do lend themselves more readily to private exploitation.

Nevertheless the returns to R & D in these products do depend upon the degree of difficulty potential competitors would have in copying the product. In some cases there are inherent technical difficulties in copying the process, or it would be prohibitively expensive, but in other cases it is the ability to obtain patent protection which creates legal barriers to potential competitors' ability to become 'free-riders' and which protects incentives to private investment in R & D.

Traditionally, however, it has been impossible for plant and animal breeders to obtain patent rights for their products, which is one reason why public-sector R & D has remained so important in this area. For, as under the European Patent Convention (EPC) of 1973, it has been judged impossible for plant varieties to satisfy one of the key criteria to qualify for a patent, namely proof of 'an inventive step'. The application of standard breeding practices to generate new varieties by crossing existing plants or animal strains has not been deemed to be invention. Thus under the EPC one set of exclusions from patentability is 'Plant or animal varieties or essentially biological processes for microbiological processes or the products thereof.'

In the absence of patent rights plant-breeding firms in particular have worked hard to obtain other means of protection and royalties for their products. The history of the development of Plant Breeders' Rights (PER) is presented by Mooney, and is of particular interest because of the concerns he expresses about the consequences of allowing the basic genetic stock, which underpins agriculture and hence our whole society, from becoming private rather than public property. While Mooney's concerns are expressed in relation to 'old' biotechnology varieties of plants, they are of particular importance with respect to the products of the 'new' biotechnology. For bio-engineered products are capable of meeting the inventive-step criterion of patentability, and both micro-organisms and transgenic animals have now been patented in the USA.

Before examining developments in the seed industry it is worth touching upon the implications of changes in research policy which stress increasing reliance upon private R & D to develop and exploit the new biotechnology and other technologies.

In the UK government has decided that it should withdraw from funding what it terms 'near-market research', that is, research beyond the basic phase and which is preparation for commercial exploitation. This policy is based in part upon the arguments that industry should be investing more heavily in R & D, and that research strategies at the near-market stage should be driven by assessments of likely commercial success which can best be made by the firms involved, and will therefore lead to greater efficiency in allocating research funds.

This has led to the closure of a number of government-financed agricultural research institutes, the scaling down of others and the sale of the National Seeds Organization by auction to Unilever; Unilever won against competition with BP and other major public companies.

This deliberate attempt to switch an increasing proportion of agricultural R & D expenditure from the public to the private sector carries with it a number of risks. In the first place there is a controversy about whether there is under-investment in R & D so that the returns to extra expenditure are high, or whether the converse is the case.

If there is under-investment, then withdrawal of public support for applied research will exacerbate this, as the private sector will only undertake R & D expenditure on those products and processes from which exclusive benefits can be captured by the investor.

Thus in relation to biotechnology in the USA, where again public support for applied biotechnology research is relatively small, Stallman and Schmid (1987) argue that there will be emphasis on technologies which are applied in factory conditions where secrecy and control can be maintained. For technologies which cannot be confined to factories, such as seeds, Stallman and Schmid state that 'Firms are also considering mechanisms which "scramble" the genome of a plant in the second generation', in order to prevent farmers from reproducing seed with the enhanced, engineered characteristics. Clearly such actions are designed to frustrate the maximum spread of benefits from the technology and to maximize private profit for companies investing in research.

Another facet of commercially orientated biotechnology research is that it will aim at the most important crops and livestock products, those produced by the largest farming units, and those which are most heavily subsidized. In the latter case this will worsen the budgetary problems of adjusting agricultural policy in OECD countries. The probable neglect of minor crops, difficult habitats and small farmers means that the public sector will have a defined role in agricultural technology research, but one in which it will be relegated to the second division and where it is unlikely to prove successful in terms of the commercial yardstick of rates of return which is increasingly emphasized by public-research policy.

TECHNOLOGICAL PROCESS IN GENETIC ENGINEERING

In order to understand how genetic manipulation is accomplished, it is important first to understand the structure of deoxyribonucleic acid, or DNA. Within its chemical structure, DNA stores the information that determines an organism's hereditary or genetic properties. DNA is made up of a linked series of units called nucleotides (Blaese), Different nucleotide sequences determine different genes genetic information. Genetic engineering is based on this genetic information.In order to understand how genetic manipulation is accomplished, it is important first to understand the structure of deoxyribonucleic acid, or DNA. Within its chemical structure, DNA stores the information that determines an organism's hereditary or genetic properties. DNA is made up of a linked series of units called nucleotides (Blaese), Different nucleotide sequences determine different genes genetic information. Genetic engineering is based on this genetic information.

Genetic manipulation is carried out through a process known as recombinant-DNA formation, or gene splicing. This procedure behind genetic engineering is one whereby segments of genetic material from one organism are transferred to another. The basis of the technique lies in the use of restriction enzymes that split DNA strands wherever certain desired secjuences of nucleotides, or specific genes, occur. This desired segment of DNA is referred to as donor DNA. The process of gene splicing results in a series of fragments of DNA, each of which express the same desired gene that can then combine with plasmids (Rubenstein).

Plasmids are small, circular molecules of DNA that are found in many bacteria. The bacteria act as vectors in the process of genetic engineering. The desired gene cannot be directly inserted into the recipient organism, or host, therefore there must be an organism that can carry the donor DNA into the host. Plasmid DNA is isolated from bacteria and its circular structure is broken by restriction enzymes (Dworkin). The desired donor DNA is then inserted in the plasmid, and the circle is resealed by ligases, which are enzymes that repair breaks in DNA strands. This reconstructed plasmid, which contains an extra gene, can be replaced in the bacteria, where it is cloned, or duplicated, in large numbers. The combined vector and donor DNA fragment constitute the recombinant-DNA molecule. Once inside a host cell, this molecule is replicated along with the host's DNA during cell division. These divisions produce a clone of identical cells, each having a copy of the recombinant-DNA molecule and thus permanently changing the genetic makeup of the host organism (Steinbrecher). Genetic engineering has been accomplished.Genetic manipulation is carried out through a process known as recombinant-DNA formation, or gene splicing. This procedure behind genetic engineering is one whereby segments of genetic material from one organism are transferred to another. The basis of the technique lies in the use of restriction enzymes that split DNA strands wherever certain desired secjuences of nucleotides, or specific genes, occur. This desired segment of DNA is referred to as donor DNA. The process of gene splicing results in a series of fragments of DNA, each of which express the same desired gene that can then combine with plasmids (Rubenstein).

Plasmids are small, circular molecules of DNA that are found in many bacteria. The bacteria act as vectors in the process of genetic engineering. The desired gene cannot be directly inserted into the recipient organism, or host, therefore there must be an organism that can carry the donor DNA into the host. Plasmid DNA is isolated from bacteria and its circular structure is broken by restriction enzymes (Dworkin). The desired donor DNA is then inserted in the plasmid, and the circle is resealed by ligases, which are enzymes that repair breaks in DNA strands. This reconstructed plasmid, which contains an extra gene, can be replaced in the bacteria, where it is cloned, or duplicated, in large numbers. The combined vector and donor DNA fragment constitute the recombinant-DNA molecule. Once inside a host cell, this molecule is replicated along with the host's DNA during cell division. These divisions produce a clone of identical cells, each having a copy of the recombinant-DNA molecule and thus permanently changing the genetic makeup of the host organism (Steinbrecher). Genetic engineering has been accomplished.

Genetic engineering is a technological process for undertaking activities which do not, and cannot, occur in nature. In this sense it is perfectly legitimate, therefore, to label genetic engineering as 'unnatural'. The European Community has incorporated this concept in its definition of genetically modified organisms, and it has been adopted by the UN Environment Programm (in the working of its Fourth Expert Panel set up under the Biodiversity Convention as part of the follow through to the United Nations Conference on the Environment and Development held in Rio de Janeiro in June 1992): 'organisms in which the genetic material has been altered in a way that does not occur naturally by mating and/or natural recombination'. Thus, hybrids and other modified organisms obtained by traditional breeding techniques are excluded from the definition of genetically modified organisms.

It is important to note that the process as well as the products are novel. The promoters of the new technology believe that a concern for the process itself is specious, downplaying its novelty whenever confronted by discussions of regulatory oversight. However, one of the bases of the decision of the US Supreme Court in the *Chakrabarti* case (which, 5 to 4, upheld the patentability of genetically engineered microbes, and which has been unquestionably accepted by foreign governments as disposative of the issue of patentability of genetically engineered life forms) was that the element of 'novelty' required by the patent law was to be found in the process by which the microbe had been created.

One reason why we cannot ignore the powerful novel aspects of the processes of genetic manipulation is because we are not omniscient as to what will, in fact, happen when we alter genomes. Genetic manipulation is not like a child's game of Legos or Tinkertoy, in which parts can be rearranged or linked up with only simple mechanical and relatively predictable consequences.

In calculating any risk from a transgenic organism, one should consider four elements: the host organism, the foreign genes, the interaction between the foreign genes and the rest of the genome, and the environment in which the organisms will be used.... [in regard to the last two elements] the literature contains many examples of genetic manipulations where inserted genes did not respond in their new environments the way they did in their old ones or where alterations with one part of the genome caused surprising activity in other parts of the genome.

There is an element of arrogance to scientific assertions which assure us that the process itself poses no risks, since scientists know so little about actual ecosystems; for example, I have been told by agronomists that over 80% of the organisms which can be identified in a soil sample from my garden are completely unknown to the scientific literature. Indeed, the Ecological Society of America itself has specifically warned about the problems presented in our lack of ecological knowledge and the resulting consequences from releasing (intentionally or accidentally) genetically altered organisms into ecological systems.

FOODS

In May 1992, the US Food and Drug Administration, responding to corporate pressures to remove the prospect of regulation of genetically altered foods, issued a set of rules which have largely left responsibility for protecting the public health and safety in the hands of the industry, permitting products to be marketed without scrutiny unless the industry indicated to the agency that it believed governmental oversight was justifiable.

However, the proposal as published by the FDA in the Federal Register perversely noted several important problem areas: creating allergens, additions of genes from sources which might violate religious and cultural norms (of vegetarians, Jews or Muslims, etc.), and implications for animal welfare and well-being (for example, the incorporation of human growth hormone-producing gene into a pig's genome produced a highly arthritic animal). Indeed, the commissioner of the FDA, David A. Kessler, and his colleagues noted the possibility that genetically engineered food might 'contain high levels of unexpected, acutely toxic substances'. In order to begin to address ethical issues presented by the genetic modification of foodstuffs, we need to have some ideas of the purposes for which these modifications are performed. (For example, if we wish to approach an analysis on utilitarian cost-benefit grounds.) Despite a great number of general statements, human nutrition and hunger do not appear to be the actual driving forces behind the development of genetically engineered foods.

Other than claiming increased shelf life (which can be indirectly related to nutrition in the sense of reducing spoilage or the consumption of foods that have begun to spoil), there is hardly any indication at all that genetic engineering is being directed to create nutritious substances out of nonnutritious ones; similarly, there are few real instances of genetic engineering reducing the cost of foodstuffs, or increasing the quantities of foods actually available to populations that are hungry because of the non-existence of consumables (as opposed to hungry because they lack money to buy sufficient foods, no matter how produced).

Indeed, early genetic manipulations seemed perversely designed to minimize the achievement of such goals: the creation of 'herbicide tolerant plants' which do not reduce the applications of dangerous chemicals to agricultural fields (which might occur if genetic manipulations of food crops were designed to make them resistant to insects, fungi, or disease) but instead permit higher levels of chemical application, or the introduction of recombinant bovine growth hormone to produce more milk at a time when developed countries suffer from milk gluts and have instituted programmes to kill cows and physically dump milk (milk as a commodity is price inelastic; quantity increases do not lead to price decreases).

The goals of genetically engineering foods seem to be more closely tied to increasing the economic gain and power of the corporations involved in food production (for example, the development of herbicide-tolerant farm crops has been led by corporations which manufacture herbicides), making production easier for corporations, reducing or altering packaging and transportation costs for the corporations, etc.

Environment

Industrial societies around the world are faced by massive pollution legacies from the previous emphases on chemical and nuclear industrial activities. The inability of the US Department of Energy to find a suitable site for a longterm nuclear waste repository illustrates our very poor track record for dealing with the scientific, technical, economic, and socio-political aspects of prior technological 'revolutions'. The environmental problems posed by genetically engineered organisms are likely to be substantially more intractable than those posed by these earlier instances of pollution because genetic wastes multiply, migrate, and mutate. A genetically engineered organism once free in the environment is impossible to recall. (Impacts which are irreversible must be considered with higher scrutiny than those which can be undone.)

Examples of environmental problems which need to be assessed include the risks of transgenic crops themselves becoming weeds; the risk of gene flow to wild relatives which might become weeds or pests; the growth of antibiotic resistance in species (particularly animals), since resistance genes are used as markers for the genetic engineering sites; the problem of exotic or non-native species taking over ecological niches (such as gypsy moth, starling, kudzu, rabbits in Australia, etc.) leading to extraordinary economic losses as well as ecological ones; the restriction (rather than increase) of biodiversity by selective advantageous breeding of transgenic organisms as opposed to natural ones, or the predatory results of expanding exotic species (such as the Dutch Elm micro-organism destroying the American Elm and severely restricting the biodiversity of certain areas in the Northeast United States); health issues (both of plant and animal species, as well as humans — worker health and safety as well as community security); and the occurrence of the completely unexpected, the inability to eliminate uncertainty (for example, the late 1993 floods in the Mississippi valley included the flooding of a field of genetically engineered corn and the dispersal of this plant material to unknown sites within the thousands of square miles of downstream flood plain).

Prospective ecological assessment is not being performed, despite the fact that its need has long since been recognized. And economic priorities, particularly those accruing directly to the promoters, as well as the nebulous spur of 'competition' with foreign countries, oftentimes overwhelms any interest in doing environmental impact analysis.

DEVELOPMENTS IN THE FOOD CROPS

Many major food crops upon which we depend originated and were first cultivated in Third World countries rather than in the industrialized countries where they are most productively exploited today and in which they have been improved by traditional plant breeding. The potato originated in South America, wheat in Ethiopia and the Near East, important maize varieties in South America, and rice in Asia. Moreover the so-called Vavilov centres, which contain the greatest variety of living species and which have been the source of much important gene material, are mainly in Third World countries.

The wealth of the industrial countries owes much to their past ability to exploit agriculturally crops and animals originating in other countries. In the colonial era the success of British scientists in smuggling rubber plants from Brazil to Sri Lanka, Singapore and Malaysia and the comparable more complex route by which coffee was introduced to Latin America from Ethiopia are instances of colonial powers obtaining significant wealth by exploiting plant material from the Third World. In one way or another the process by which institutions and firms in developed countries have continued to collect new plant and animal varieties from less-developed countries has continued. But what is still a major source of friction is the procedures by which legislative protection is developed to give companies in the richer countries commercial rights to exclusive exploitation of varieties developed from plants freely collected from other countries, and to extract high rates of profit from the sale of those varieties. Most extreme is the situation whereby under the USA Plant Patent Act of 1930, which covered asexually propagated plants such as certain fruits, flowers, ornamental shrubs and trees, it has been possible for plants discovered and smuggled out of other countries to be given patent protection in the USA.

The key to this is that the Act accepted the unfeasibility of requiring proof of an 'inventive-step' for the granting of a patent, and employed the criterion that as far as can be determined the plant variety is new. Since some novel variety found in the rain forests of South America may satisfy this criterion, and the source of the plant can be concealed from the authorities, the theft and smuggling of plants can be said to have been encouraged.

Seed companies in developed countries have successfully pressed their case for protection, and have progressively extended Plant Breeders' Rights (PBR) through both national legislation and the Union for the Protection of New Varieties of Plant (UPOV), which is an international agreement among signatory countries to honour a system of varietal rights. In the case of food crops, in the absence of patentability, PBR operates by registering distinct varieties. Pressure has been exerted upon less-developed countries to become signatories of UPOV and thus acknowledge the rights of breeders, most of whom are from developed countries. However, the majority have stalled or rejected membership, and have been fundamentally unhappy about accepting the principle of creating monopoly rights for others over plant material which in some cases originated in their own country.

The whole issue of intellectual property rights has been raised to a prominent political level by the USA's insistence that it be included as one of the thirteen areas for negotiation in the Uruguay Round of multilateral bargaining of the General Agreement on Trade and Tariffs. While the issue is a much broader one than that of plant and animal breeders' rights, it reflects a basic political conflict between the richer countries, determined to obtain secure returns on R & D expenditures by companies, and LDCs struggling to catch up and anxious to avoid having to pay royalties on products protected by PBR and patents. This conflict has broader moral and ethical dimensions when it relates to the issue of charging royalties for the seeds of food crops to countries where malnutrition is widespread.

Against the previous argument has to be balanced the need of those incurring R&D expenditure to capture enough of the returns to provide an incentive for R & D. At least that is certainly the case where R & D is to be undertaken by the private sector, which is politically the increasingly preferred solution in the UK, the USA and elsewhere. This, however, leads to another area of concern, which is that a small number of large multinational chemical and pharmaceutical firms have increasingly come to control the seed industry. Companies such as Shell, Sandoz, Dekallb-Pfizer, Ciba-Geigy and BP have become owners of most of what were small independent seed companies.

This linking between chemicals and seeds within firms has a strong commercial logic. It facilitates the marketing of packages consisting of seed varieties plus the appropri-ate fertilizers, insecticides, etc. It also allows the companies concerned to integrate their plant breeding, biotechnology and chemical research. One particular synergy for these companies will be to bioengineer varieties of major food crops to resist their own herbicides and weedkillers. This would permit chemical weed control to be extended to large-scale crops where this is not now possible, and is just one way in which the direction of biotechnolo-gical research may be influenced (biased) by the combination of forces giving a small number of companies extensive control over the basic means of agricultural production.

There are many other issues which could be touched on regarding corporate influence over the development of agricultural biotechnology. It is, however, understandable that the spokesmen for LDC countries, which are profoundly concerned about their dependence upon western technology and exercised by what is seen as colonial and post-colonial exploitation by the West, should be alarmed by the prospect of new rights being created over seeds and other biotechnological products. Certainly there are widespread fears that they will only gain access to the fruits of biotechnology on unfavourable terms and that their relative dependence on and subservience to western industry will be increased so that benefits to them will be small. There are many western critics who would agree with this, and that the system by which agricultural biotechnology is delivered will favour the 'haves' rather than the 'have-nots'.

AGRICULTURAL BIOTECHNOLOGY: THEIR IMPACT

There are divergent views about the rate of uptake of new agricultural biotechnologies and their impact. Kalter and Tauer and the US Office of Technology Assessment anticipate comparatively rapid take-up, while others (Buckwell and Moxey and Farrington do not foresee much impact until the next century, that is until ten to fifteen years have elapsed. The caution of the latter commentators is probably justified, for there are various hurdles to be jumped before biotechnologies can gain acceptance.

One hurdle to be overcome is the economic one. Self-evidently there must be some economic advantage to farmers or agro-industry from adopting the technology. Frequently what captures the scientific imagination turns out to be economically unviable in use, although it is only with use that efficiency can be improved and costs brought down.

A more demanding test still for biotechnologies will be to obtain legislative approval and public acceptance. These are intimately connected. In the case of machinery developments, provided safety standards are observed there are no legal impediments to developing new and improved machines, since there are no obvious problems of hazard to the public and therefore public acceptance. Chemical insecticide, fungicides and herbicides do, however, pose much greater problems because of concerns with toxicity, and elabourate testing and licensing procedures have evolved after a relatively haphazard set of procedures in the 1950s and 1960s, when widespread use was made of DDT and organo-phosphorous compounds without adequate recognition of their toxicity. In significant part, because of the problems which have been experienced with agro-chemicals, the licensing of biotechnologies will inevitably be based on testing at least as stringent as that which now exists for chemicals. Indeed the licensing procedures are likely to be more stringent because of public and scientific concerns.

Public acceptance of biotechnology in the food chain will be made more difficult by confusion about differences between biotechnologies. In the early 1980s there was a crescendo of concern about the use of steroids to promote faster liveweight gain in calves and cattle. The public outcry eventually led to the banning of such hormones for meat animals; but the legacy is a profound distrust of and hostility to new products such as synthetic Bovine Somatrophin (BST), which has the capacity to increase the milk yield of those dairy cows which are 'relatively' deficient in what is a naturally occurring hormone.

Although initial scientific results are favourable, and some minor doubts remain, the major influence leading the European Commission to impose a moratorium on the use of BST until the end of 1990 at least is concern about public perception; the Commission has stated 'It would be a serious setback to producers and to the Community's milk policy were present [positive] trends in consumption to be reversed as a result of adverse consumer reaction.

With newly bioengineered plants and animals scientific and public concerns are emerging which are of a different order from those associated with previous biotechnology. One relates to the possibility that crop failures may become more frequent if biotechnology leads to a reduction in genetic diversity in crops being grown; such a tendency has already been observed in relation to the Green Revolution technology for wheat.

There are also concerns about upsetting the balance of nature in unforseen ways, akin to the unanticipated consequence of introducing rabbits into Australia and then trying to control these by introducing myxomatosis. Thus questions arise such as what happens if herbicide resistance introduced into a commercial crop plant transfers itself by cross pollination into a closely related weed species, or indeed if the herbicide-resistant crop should colonize wild habitats. There are, of course, more lurid and absurd notions bandied about in the popular press which are similar in nature to the attacks made on

Darwinism. The questions, both absurd and real, raised in relation to agricultural biotechnology will undoubtedly slow the rate at which it is adopted. Nevertheless instances of adoption are increasing: BST in the USA, a bioengineered baker's yeast in the UK, bioengineered sheep producing insulin, etc. As these become more widespread and increasingly affect lower-valued, bulk agricultural products, so the impacts of biotechnology on structural change will increase. The balance of economic power will switch increasingly to industry, to high-technology large farms and against smaller farmers in disadvantaged regions and countries. That, unfortunately, is a seemingly inevitable consequence of what we consider to be economic progress.

PLANT PESTS AGENTS

A pest is "a plant or animal detrimental to humans or human concerns (as agriculture or livestock production)"; alternative meanings include organisms that cause nuisance and epidemic disease associated with high mortality (specifically: plague). In its broadest sense, a pest is a competitor of humanity. 'Pest' is a generic word for other terms such as: insect pests (of agriculture or human disease vectors), vermin, weeds and plant pathogens. In the past, the term might have been used for detrimental animals only, thus for example, causing confusion where the generic term 'pesticide' meant 'insecticide' to some people.

A plant pest is any agent, substance or organism, or parts thereof, that may cause injury, disease or damage, either to plants or to the environment in which it is introduced. APHIS maintains a list of organisms considered to be plant pests, and therefore, subject to regulation under the FPPA. If an organism is not on the list, it may still be subject to regulation as a plant pest if there is reason to believe it is or will act as a plant pest. Organisms considered plant pests, or that exhibit the potential to be plant pests, are treated as 'regulated articles' under the FPPA.

The regulations extend to the introduction of GM organisms meeting the definition of a plant pest or for which there is reason to believe they are plant pests. APHIS considers the environmental release of a GM crop to be equivalent to the introduction of new organisms. Therefore, until proven otherwise, a GM crop is considered a 'regulated article' under the FPPA. The use of the term 'plant pest' in reference to GM plants means the 'non-pest' nature of the crop has yet to be determined. To date, all field tests conducted have demonstrated that GM crops exhibit no more 'pest-like' traits than their conventional counterparts.

All GM plants developed to date have involved the use of at least one designated 'plant pest' as either promoters or vectors, and as a result, have been subject to regulatory review under the FPPA. APHIS takes the position that an entire plant can be designated as a regulated article, even if it was not developed using plant pests, if the plant is the product of genetic engineering, and the agency determines or has reason to believe it is a plant pest. Given this position, APHIS would likely challenge any attempt to introduce a GM plant into commerce unless the developer has first gone through the APHIS regulatory review process.

REQUIREMENTS FOR COMMERCIALIZATION

A number of years of field testing are required to evaluate the agronomic and product quality characteristics exhibited by new plant varieties developed within a labouratory or greenhouse. APHIS regulations outline procedures for obtaining a permit or providing notification prior to the importation, interstate movement or release of a regulated article, which of course, includes field testing.

Prior to an introduction, a proponent must provide notification to APHIS of its intentions or submit an application to obtain a permit. The notification and permit requirements have evolved as an extension of the long-standing programm for the regulation of plant pests within the United States. Since 1987, APHIS has issued permits or acknowledged the receipt of notifications for the field release of more than 6500 new crop varieties.

Permits

For APHIS approval, developers must demonstrate that a new plant variety poses no significant risk to other plants in the environment or is at least as safe as similar crop varieties. APHIS review and approval are required for the shipment of seed and for the conduct of field trials involving new crop varieties. APHIS issues site-specific permits for field tests or releases into the environment.

Prior to field testing, APHIS may seek further clarification of study objectives, specify how, when or where research may be conducted, establish the data to be generated, collected and reported, stipulate additional requirements for the monitoring or securing of test sites, and specify methods for the disposal of crop residues that remain at the end of a study. The results of field trials are often used in making a determination as to whether a GM plant exhibits any effects on non-target species, or poses any unique plant pest problems.

For field trials conducted under either notification or permit, developers of GM plant varieties are required to submit reports to APHIS that include: details regarding the methods of observation; the data collected; and interpretations concerning the effects on plants, non-target organisms, or the environment. In the event that a GM crop is likely to be the subject of a petition for a determination of non-regulated status, developers must provide a description of known and potential differences, and substantiate that the regulated article is not likely to pose a greater plant risk than the organism from which it was derived.

An environmental assessment is a requirement of the NEPA, Council on Environmental Quality regulations, and USDA procedures for issuing permits under the FPPA. Under the permit process, developers of GM plant varieties must disclose information concerning the plant, test facilities, and control measures in place for its transport and field testing. Based on this information, the USDA performs an assessment of the potential environmental impact following a release. If the agency reaches a 'Finding of No Significant Impact' (FONSI), a permit may then be issued.

Notifications

Depending on the plant species and intended introduction, applying for a permit from APHIS can be a complicated and time-consuming process. In the spring of 1993, APHIS introduced a simplified notification process as an alternative to applying for a permit, which in most instances applies to the introduction of GM plants. To qualify for notification, a GM plant must be introduced in accordance with the eligibility criteria.

Originally, six GM crops (corn, cotton, potatoes, soybean, tobacco and tomatoes) qualified for notification; however, the process was more recently expanded to include most other crop species not capable of becoming a noxious weed. The developers of new GM plant varieties are encouraged to contact APHIS when making a determination as to whether a GM plant would qualify for notification. If a particular GM plant does not qualify for notification, a more involved regular permit process must be followed.

For a GM plant to qualify for notification, it and its predecessor must not be a noxious weed, the introduced gene sequence(s) must be stably integrated within the host, not originate from a plant or animal pathogen, exhibit a known function that will contribute to plant disease, or encode for substances that could be potentially toxic or infective to non-target organisms, such as pharmaceuticals, or viral components other than those coat proteins already established as safe.

Performance standards also exist for the conduct of field trials to ensure that an adequate level of containment is provided for all introductions. It is the responsibility of the developer to ensure that an adequate level of containment is provided for field trials involving GM plants under notification. The performance standards utilized will vary according to the biology of the plant and the nature of the introduction. In general, this includes specifying methods for the proper handling, shipment, field monitoring and disposal of plant material at the end of the study.

As a part of the notification process, an applicant must provide information about the plant, identify the source of any genes used; the method of genetic modification; and the size, date and location of any proposed field test or introduction. Notification is required at least 10 days prior to the interstate movement or 30 days in advance of the field testing or importation of a regulated article that qualifies for notification with APHIS.

Determination of Non-regulated Status

Once sufficient data have accumulated through labouratory and field trials performed under permit or notification, a developer may petition APHIS for a determination of non-regulated status. A DONRS allows a GM plant to be grown, tested or used for crop breeding without further regulatory oversight by APHIS. APHIS will grant a DONRS if sufficient evidence can be provided to demonstrate that no plant pest potential exists (*i.e.* that the GM plant is not expected to become a pest, poses no significant risk to the environment, or is as safe as conventional plant varieties).

The developer is required to demonstrate a lack of change in disease or pest resistance status, an absence of any potential for contributing to development of a new plant pathogen or pest, as well as to address any questions relating to potential environmental consequences of the release.

APHIS examines several parameters when making a determination as to whether a GM crop should be considered a plant pest. Reporting requirements include submitting a detailed rationale for development; overview of the biology of the crop (competitiveness, survivability, dormancy); molecular characterization; protein expression; morphologi-cal/phenotypic characteristics; outcrossing/geneflow; weediness; insect and disease susceptibility; and, recombination potential (for viral genes).

Other considerations include assessing whether a GM crop could itself become a weed. In making this determination, factors that are considered include: how the seeds are dispersed in the environment; whether the seeds are capable of surviving over the winter; the potential for the development of volunteer plants; and whether these plants could reproduce into viable offspring. Another important consideration is whether the seeds or plants are capable of surviving outside of a managed agricultural environment (*e.g.* watering, fertilizer, etc.), and how the introduced traits could influence the viability of the crop under marginal conditions.

APHIS also takes into consideration potential adverse effects on wildlife, including birds, beneficial insects and mammals, following the introduction of a GM crop. Some of this information comes from field trials that are conducted in multiple locations for several years. By observing crops growing under actual conditions of use in the field, scientists can compare insect populations existing within a field planted with GM crops to those coexisting in fields planted with a non-modified variety of the same plant. Field trials are also useful in identifying any changes in the plant physiology, such as height, colour, leaf placement, time of flowering, etc., that result from modifications in the genome of the plant.

Monitoring of these changes during field trials allows APHIS to evaluate how these changes might benefit or adversely impact the behaviour of wildlife in contact with these crops. Knowing that wildlife, such as deer, often feed on agricultural crops, APHIS also takes into consideration the nutritional content of the GM crop. Comparisons of the essential nutrient content are often made between the GM plant and its conventional counterpart. Other interests may include determining the impact of the modification on the levels of adverse factors (*e.g.* natural toxicants or anti-nutrients) present in the plant.

APHIS maintains a list of all crops that have received a DONRS. At any time, APHIS retains the authority to prohibit the commercial planting of a GM crop if it is determined that it is becoming a plant pest.

PETITION PROCESS

Before a GM plant can be freely transported and commercialized, the developer must first petition the APHIS for a DONRS. A petition for a DONRS requires the submission of extensive data on the introduced gene construct, effects on plant biology and effects on the ecosystem, including the spread of the gene to other crops or wild relatives. Developing all data necessary to support a petition for a DONRS may take months or years depending upon the scientific issues that need to be addressed.

Based on the information provided within the petition, APHIS evaluates the potential impact (*e.g.* toxicological) that the introduced modifications will have on non-target organisms in the environment, including threatened or endangered species. A petition for a DONRS must provide a rationale for the development and introduction of the GM crop, starting with genotypic and phenotypic information on both the host and donor organism(s).

This includes a detailed description of the methods of transformation, identifying all gene sequences (*e.g.* promoters, leader sequences, introns, selectable markers, etc.), their function, and potential for plant risk following introduction. A combination of different analytical methods may be used to demonstrate the stable integration of an introduced gene sequence within the genome, its inheritance and expression in host plants. To the extent possible, the gene(s) of interest (*e.g.* resistance, marker genes), levels of expression, and im pacts on levels of inherent plant toxicants in plant tissues, under experimental and actual growth conditions in the field, must be characterized.

Field performance studies (*e.g.* leaf morphology, pollen viability, seed germination, viability, insect susceptibilities, disease resistance, yield, etc.) are used to identify any differences or similarities in phenotypic characteristics exhibited by the GM and non-modified crop. The extent of field performance data required in support of a petition for non-regulated status depends not only on the nature of introduced gene sequences, but also on the physiology of the host plant. Less extensive field performance studies are required for plants that are: highly domesticated (*e.g.* corn); self-pollinating (*e.g.* soybean); male sterile; those with high seed germination rates (>90%); or unlikely to influence their potential weediness or fitness (*e.g.* delayed ripening, oil seed content). Whereas, GM crops exhibiting a tolerance or resistance to cold, salt, biotic (*e.g.* insects, pathogenic agents), or other abiotic (*e.g.* herbicide) stresses, are subject to more extensive field performance testing.

APHIS considers several parameters in reaching a DONRS, including whether the GM plant can cross-pollinate with other plants in the wild, and if it can, the ecological consequences (*e.g.* insect resistance, herbicide tolerance) that might result. Extensive databases are maintained on species capable of cross-pollinating with GM crops (*e.g.* corn, soybean, cotton, canola, etc.), based on the results of years of breeding experiments, biological surveys and research conducted by agricultural and weed specialists within APHIS. In instances where out-crossing might be expected to occur, APHIS may stipulate the planting of refuge areas to limit the potential for adverse ecological effects following commercial planting.

AGENCY RESPONSE

APHIS will grant a determination of non-regulated status if the developer can sufficiently demonstrate a lack of plant risk. In making a determination that a GM plant will no longer be regulated, APHIS prepares two documents, an environmental assessment and a DONRS to satisfy regulatory requirements under NEPA and FPPA, respectively.

Both documents are developed based on the review of data submitted as a part of the petition for a DONRS. Once the DONRS has been granted by APHIS, a GM crop no longer requires notification or permit prior to its movement or release within the United States.

A review for DONRS is generally completed within 10 months of submitting a formal application to the APHIS, allowing sufficient time for publication in the Federal Register, and opportunity for public comment. Notices of all petitions for a DONRS are published in the Federal Register, and the public is given 60 days to comment for or against the petition, after which APHIS has up to an additional 180 days to either approve or deny it. APHIS provides access to a considerable amount of information relating to approved field trials and petitions received for GM crops. The petition for a DONRS must include scientific details regarding the genetics of the plant, the nature and origin of the genetic material used, the potential for indirect effects on other plants, and any other information that could be considered unfavourable to the petition.

Extensions of Non-regulated Status

In certain instances, the non-regulated status of a GM crop may be extended to include other varieties of the same crop species, provided that changes in introduced gene sequences are insignificant, and no new plant pest issues would be expected. For an extension, developers must substantiate that the new regulated article poses no serious issues meriting review under separate petition, by providing a precise description of the genetic modifications to the regulated article and a detailed comparison to modifications made in the GM plant previously granted non-regulated status by APHIS. Field trials must be performed under notification or permit to demonstrate the similarities in the phenotypic properties. In support of multiple extensions, product identity standards may be submitted describing the genotypic and phenotypic properties exhibited by new plant varieties considered extensions of an existing DONRS for a GM crop.

Future Considerations

The USDA recently appointed an Advisory Committee for Agricultural Biotechnology that will work in conjunction with the NAS (Standing Committee on Biotechnology Food and Fibre Production, and the Environment) on the completion of a critical review of the regulations and policies used by APHIS in the commercial approval of GM crops. The committee comprises scientists, farmers and representatives of consumer groups and seed companies, and is anticipated to take two years to complete its assignment.

In addition to APHIS, other functional areas within the USDA are beginning to get more actively involved with the regulation of agricultural biotechnology. For example, the Grain Inspection Packers and Stockyards Administration (GIPSA) has made a commitment to work with farmers and industry groups in the development and validation of reliable test methods and quality assurance programms to differentiate GM from non-GM commodities.

As a part of this commitment, GIPSA intends to publish a proposed rule on the standardization of test methods and approaches for the detection and identity preservation of grains. GIPSA is also intending to offer accreditation services for labouratories testing grains for the presence of GM content, and to evaluate commercial test kits to ensure they can be considered accurate and reliable. It is expected that an administrative fee will apply to the accreditation services being offered by GIPSA.

The USDA also has made continuous commitments to funding research that will assist federal regulatory agencies in making science-based decisions concerning the safe introduction of GM organisms into the environment. Active areas of research include assessing the potential risks associated with the environmental introduction of GM plants (*e.g.* gene flow), the cumulative effects of large-scale commercial plantings, the potential interactions between GM and non-GM crop species, programmed resistance, and the development of statistical methodology and quantitative approaches to measuring the risks associated with the field testing of GM plants, etc.

Environmental Protection Agency

The EPA has authority over the registration of chemical and biological pesticides under the Federal Insecticide, Fungicide and Rodenticide Act (FIFRA). A pesticide is any product 'intended for preventing, destroying, repelling, or mitigating any pest'. All pesticides require registration by the US EPA prior to their distribution. As a part of the registration process, an applicant is required to submit information and data in support of the safety of a pesticide in its intended use. In addition, US EPA is responsible for establishing acceptable tolerance levels for registered pesticides on or in raw agricultural food commodities under the FFDCA. At any point, US EPA has authority to amend or revoke a registration or residue tolerance, if an adverse effect is observed, or if the risks associated with the use of the pesticide are determined to be unacceptable.

Role

Under the Coordinated Framework, the US EPA was identified as being responsible for the review, assessment and registration of the products of rDNA technology that act as pesticides. Leading up to this publication, the US EPA had acquired considerable expertise in the registration of biological pesticides (*e.g.* microbial). The US EPA was first involved in discussions regarding the potential risks of pesticides developed using rDNA technology in the early 1980s, during a series of public meetings involving the FIFRA Scientific Advisory Panel (SAP) and the Biotechnology Scientific Advisory Committee (BSAC).

It was recognized that the potential risks posed by plants modified to express pesticidal components through rDNA technology were unique and warranted additional consideration relative to conventional pesticides. For example, the potential for introduced genetic material to out-cross from a modified plant to a sexually compatible wild or weedy relative through pollen spread was one of the unique, theoretical risks posed by plant-pesticides.

As a result, the EPA proposed a plant-pesticide rule in 1994, outlining their interests in the regulation of products. A package of three proposed final rules was released early in 2001, redefining regulations for plant-incorporated protectants (PIPS), previously referred to as plant-pesticides. Essential elements of the proposed rules for the regulation of GM plants expressing pesticidal traits have been followed since 1994. The final rules provide further clarification of how PIPS will be regulated under FIFRA and FFDCA, including specific exemptions and became effective in September 2001.

Definition of Plant-incorporated Protectants (PIPS) or Plant-pesticides

US EPA's final rule clarifies the regulation of whole plants that contain PIPS. When plants are used intentionally for controlling pests in some way, such uses meet the definition of a pesticide under FIFRA. Nevertheless, the agency exempts the plants themselves from regulation, focussing instead on the substances they contain, previously referred to as plant-pesticides and now known as PIPS. A plant-incorporated protectant (PIP) includes the substance produced by the living plant, and all genetic material necessary for its production (including promoters, enhancers, etc.).

Thus, genes and substances (*e.g.* enzymes) responsible for the conversion of natural plant constituents into pesticidal substances would be considered PIPS. To date, PIPS registered by the US EPA have been proteins and the genes required to make these proteins within the plant. Crops involved include potatoes, cotton, field corn, sweet corn and popcorn.

Clearly, Bt toxins represent the most predominant class of PIPS that have been registered by the US EPA. Major reasons for this include the success of microbial products containing Bt and their long history of use for nearly 40 years. It has been estimated that combined, Bt corn, potato and cotton were cultivated on approximately 10 million acres in 1997, 20 million in 1998, and 29 million in 1999. As a result, less chemical insecticide has been applied, and there have been significantly higher crop yields. In addition, due to reduced opportunity for opportunistic fungal infections of insect-damaged corn, lower levels of mycotoxins have been produced, reducing the toxicological risk to both humans.

Exemptions

US EPA has concluded that living plants with pesticidal activity do not require a high level of regulatory scrutiny by the agency and thus, are exempt from regulation under FIFRA. The agency commented that focussing on the substance produced within the plant, *i.e.* the PIP eliminates any need for the US EPA to register plants, thus conserving agency resources and reducing potential overlap with other agencies, notably the USDA. For example, although whole plants that are used as biological control agents themselves, such as chrysanthemums, are exempt, substances that are extracted from plants, such as the insecticidal material pyrethrum, are not excluded from regulation under FIFRA.

Further exemptions have been proposed, based on familiarity and presence of the pesticidal substances in the food supply. Only one of the additional proposed exemptions has been included in the final rule, and all others have been proposed for further public comment. Within the final rule, EPA has exempted PIPS and encoding genetic material originating from plants sexually compatible through conventional breeding but not through rDNA technology. The rationale for this exemption is that such substances could be bred, either naturally or through conventional plant breeding, from close plant relatives that share genetic material from a common gene pool.

On the other hand, genetic modifications that were never before possible can be made using rDNA technology, involving the genes from sexually incompatible plant species. Therefore, the exemption of plant-pesticides developed using rDNA technology, involving genes from sexually compatible plants has not been finalized as a part of the supplementary proposal to the final rules for PIPS.

Other exemptions for further comment have also been identified within the supplemental proposal. The first group are PIPS that 'act primarily by affecting the plant'. The group have in common some sort of defence mechanism such that a pest would be less able to attach to, penetrate, or successfully invade a plant. Examples include:

- Some sort of structural barrier such as wax or plant hairs;
- Inactivation of or resistance to a pest or substance (*e.g.* toxin) produced by a pest;
- Creation of a deficiency in a nutrient or growth factor required by a pest;
- Hypersensitivity response that would contain or limit spread of the pest (*e.g.* necrosis of surrounding plant tissue, creating a functional barrier);
- Plant hormones that are naturally occurring but with change can affect a plant in a variety of ways with potentially dramatic results.

Herbicide-tolerant Crops

In the case of herbicide-tolerant crops, EPA determines whether the property of increased resistance may encourage higher rates or more extensive application of an agricultural chemical to food crops. The higher exposures to herbicides would be subject to evaluation through risk assessment, requiring the submission and review of detailed data. If the risk is determined to be acceptable, a herbicide product would require label extensions for the addition of new tolerant crop varieties.

Experimental Use Permit (EUP)

Small-scale research involving PIPS is covered under the permit and notification process that exists for new plant varieties administered by APHIS. The US EPA requires that an Experimental Use Permit (EUP) be obtained for all research and field testing involving greater than 10 acres of a plant modified to express a pesticidal substance. The EUP process allows a developer to gather the product use, performance, residue and other types of data, required in support of the registration of PIPS. All US EPA decisions concerning applications for EUPs are published within the Federal Register for comment.

Registration requirements

For the registration of a plant-incorporated protectant, the EPA evaluates the potential impact, on human health and the environment, of the pesticide substance and of the genetic material necessary for its expression within the modified plant. With this in mind, the EPA developed their guidance for evaluating the safety of PIPS based on data requirements that were established for microbial pesticides.

A number of modifications to the reporting requirements were required to focus the assessment on non-target toxicity rather than pathogenicity, as well as to evaluate the expression of the pesticide substance within the plant itself, rather than by spraying, etc. Although specific reporting requirements for the registration of PIPS are established on a case-by-case basis by the EPA, the general categories of information and data required in support of the registration of a plant-pesticide include:

- Product characterization
- Toxicology
- Environmental fate and exposure
- Non-target organism effects
- If necessary, plans for pest resistance management.

The EPA continues to hold meetings of their SAP to re-evaluate the reporting requirements that have been established for PIPS and to ensure that all potential human health and environmental concerns are being addressed as new scientific information becomes available. In addition, the developers of PIPS are encouraged to consult with US EPA during all phases of the research and development process, particularly when it concerns an environmental release.

Product Characterization

Product characterization allows for consideration of any potential hazard(s) that might be associated with anticipated exposure to a plant-incorporated protectant. This process includes a review of the source of introduced genetic material, including all regulatory sequences, the modifications that have been made, methods of transformation, inheritance, stability and expression in the host plant.

For the introduced pesticidal substance, the anticipated modes of action, specificity and toxicity to both target and non-target organisms are evaluated. Target organisms may include weeds, insects and diseases (viral, bacterial, and fungal) affecting a crop. The crop itself, as a volunteer weed, may also be of concern following crop rotations. Once a PIP and any potential for exposure have been characterized, the data required in support of an environmental and human health risk assessment can be defined.

Ecological Effects

Environmental fate and exposure are important considerations when evaluating the potential impact of PIPS on the environment or on non-target organisms (*e.g.* wildlife, beneficial insects). Depending on the crop and introduced pesticide substance, data on the level of expression and rate of degradation may be required, first for the tissues of the modified plant, and secondly, the fate in soil, water, or any other environmental compartment. Such data allow the potential for exposure to the pesticide substance to be evaluated for non-target organisms via feeding on, or through decomposition of, the modified plants.

Studies are arranged in tiers, starting with Tier I, consisting of acute toxicology and basic characterization studies that help predict environmental fate. Depending on the results obtained in Tier I, and the level of concern, additional studies may be requested from higher tiers. It is expected that most PIPS will only undergo Tier I testing, since negative results are expected at relatively high exposure levels.

The 'maximum hazard approach' when used in Tier I provides a high level of assurance that adverse effects would not be expected to occur under actual conditions of use within the field. If the results of Tier I reveal a potential hazard at doses within the range of expected concentrations in the environment, additional studies would be requested in higher tiers to better characterize the potential for any kind of effect.

Data requirements may be satisfied by generating test data or by requests for data waivers with credible justification. The type and extent of dietary studies required will depend on properties of the crop and on the introduced pesticide substance. Dietary studies typically involve feeding plant tissues (*e.g.* grain, pollen) that express the PIP or that have been spiked with the pesticidal substance.

However, when determined necessary, the pure pesticidal substance may be administered to non-target species at doses of up to 10 to 100 times the expected level of exposure in the field. Dependent upon the potential for exposure, additional studies may also be requested involving any number of other non-target organisms (*e.g.* honeybees, green lacewing, ladybird beetles, parasitic wasps, earthworms, etc.). When toxicity is observed, the potential for exposure becomes an important consideration when ascertaining whether an adverse effect might be expected under actual conditions of use in the field.

The requirement for provision of a resistance management plan is not routine practise for PIPS, but depends on an evaluation of the potential for resistance to develop and the threat to existing or functionally equivalent pest management practises.

To date, resistance management plans have only been placed on Bt PIPS. If considered appropriate, the EPA may request that a resistance management plan be submitted as a part of a condition for registration. A resistance management plan may require a commitment towards the generation of additional research data, to implement structured refuges, to perform annual resistance monitoring, to initiate remedial action plans, and to educate growers. The EPA has worked closely with academia, other federal agencies, public interest groups, industry, and growers, to establish resistance management plans based on the most current science.

Human Health Effects

For agricultural food crops, special consideration is given to evaluating the safety of the pesticidal substance introduced as a component of consumed foods. Rodents (often mice, to conserve test material) are orally administered the pesticide substance at doses of up to 100 000 times the levels of expected human consumption as a part of the diet.

Although the modified plant tissues would be considered the most appropriate vehicle for determining the acute toxicity of a PIP, it is not typically feasible, since only small amounts of the pesticidal substance are present in plant materials. Therefore, an alternative source (*e.g.* microbial) is often used to generate a sufficient amount of identical pesticide substance for testing purposes. In this case, the equivalency of the test substance obtained from an alternative source to the PIP (*e.g.* composition, identity, etc.) being expressed within the plant is an important consideration that must be well documented.

For pesticide substances that are proteins, the principal concern relates to the potential for the introduction of a new allergen or toxin into the food supply. Since most allergenic food proteins are stable to digestion, *in vitro* digestibility studies are used to predict how long a pesticidal substance may persist in digestive fluids (*e.g.* gastric, intestinal) following ingestion. In addition, heat stability studies are performed to provide an indication of the fate of the PIP during food processing.

Commonly, assays of amino acid sequence homology are performed to determine similarity to substances that are known to pose unique safety concerns (*e.g.* toxins, allergens). The EPA uses a 'decision tree' approach similar to that followed by the FDA in assessing the potential allergenicity of a new food protein. In many instances, US EPA and US FDA collabourate with each other when evaluating issues of safety for new pesticidal proteins being expressed in GM food crops.

Residue Tolerances for Plant-incorporated Protectants

Analogous to the consideration of tolerances for residues of conventional pesticides, the US EPA reviews data on the human, animal and environmental safety of PIPS to determine whether residue tolerances should be established for the amounts expected to be present in foods derived from the GM plant. Under Section 408 of the FFDCA, if the GM plant expressing a PIP is a food crop, the US EPA is obligated to establish the 'safe level' of pesticide residue allowed, or tolerance level.

The Food Quality Protection Act (FQPA) served to amend the FFDCA and FIFRA by outlining the process by which the safety of pesticide residues on raw or processed foods can be established. Under this process, in order for a pesticide residue to be considered safe, a 'reasonable certainty of no harm' must exist assuming aggregate exposure of the public, taking into consideration sensitive children within the population. To date, all pesticidal proteins that have been registered as PIPS have been determined to be non-toxic, and as a result, have been considered exempt from the requirement to establish a tolerance under the FFDCA.

Consistent with categorical exemptions as originally proposed for plant-pesticides under FIFRA, the US EPA also proposed that tolerances might not be required for pesticidal substances derived from sexually compatible plants, regardless of the process used to introduce the PIP, provided that the genetic material encoding the pesticidal substance was from a plant commonly used as food, and that the presence of the pesticidal substance in a host plant did not result in any new or significantly different dietary exposures. Final rules include exemptions from tolerances for the residues of PIPS, and their encoding nucleic acids, but only for PIPS derived through the conventional breeding of sexually compatible plants. US EPA is requesting further comment on the possible exemption from the requirement of a tolerance for:

- PIPS derived through rDNA technology;
- PIPS that act by affecting plant defences;
- PIPS developed based on viral coat proteins.

Future Considerations

The US EPA anticipates that the requirements for the registration of PIPS will continue to evolve as the science and policies relating to biotechnology continue to mature. In light of on-going criticisms that the plant-pesticide rule has received since its publication in 1994, a number of stakeholder workshops have been hosted by the US EPA. In March 1999, the House Agricultural Subcommittee hosted a congressional hearing to address a number of these concerns. One of the primary questions addressed was whether the definition of a pesticide under FIFRA provides the US EPA with the authority to regulate pest-protected plants that have been developed using rDNA technology.

Another concern was that the proposed rule was too broad in scope because all plants have natural defence mechanisms, and as such, would be classified as 'plant-pesticides'. In addition, an exemption from the plant-pesticide rule for those plants not developed using rDNA technology would serve to regulate the process used in the development of the GM plant, rather than on the pesticidal substance being produced within the plant. Further criticism related to the use of the term 'plant-pesticide' in relation to the products of the genes that had been introduced into GM crops.

As a part of its final rule, the US EPA recommended that plant-pesticides be renamed PIPS, an alternative name which distinguishes them from the other types of pesticide products requiring registration with the US EPA. A new FIFRA section has been added to the Code of Federal Regulations specific to the regulation of PIPS. In doing so, the US EPA has acknowledged that the new final rule would be making a distinction between the products of conventional breeding and rDNA technology, *i.e.* regulating on the basis of process not product. One reason given for this divergence was that the use of this criterion would be expected to provide the public with increased confidence that an appropriate level of regulatory oversight was in place for assessing the risks of rDNA technology, which is a societal not scientific issue.

SAPs are charged with providing the Office of Pesticide Programms (OPP) of the US EPA with advice and guidance on human health and environmental issues relating to proposed regulatory decisions for pesticides, including PIPS, based on the latest scientific data. For example, the SAP for Bt Plant Pesticides met three times in the year 2000 to consider issues related to the safety of plants that had been genetically modified to express insect-specific Bt toxins.

The US EPA released a decision to temporarily extend registrations for all currently registered pest protected plants throughout the 2001 growing season. Additional provisions, such as the recently strengthened resistance management requirements for plant-pesticides, along with the original registration conditions, were considered to provide adequate protection during the extended time period. In addition, the reassessment process for these products has been designed to assure maximum transparency and opportunity for public comment in the regulatory decision-making process.

CONCLUDING COMMENTS

The US FDA has been praised for its science-based regulatory approach to evaluating the safety of products developed using rDNA technology based on the characteristics of the food product, and not the means by which it has been created. On the other hand, the regulatory approach of both the USDA and US EPA has been criticized for discriminating against the products of rDNA technology.

A consortium of 11 major international scientific societies and the Council on Agricultural Science and Technology (CAST), an organization representing 36 scientific and professional groups, have strongly criticized the US EPA for their proposed approach to regulating plant-pesticides on the basis of the process used in its development rather than the characteristics exhibited by the plant. More recently, the US Congressional Committee on Biotechnology released a report entitled 'Seeds of Opportunity' recommending revisions to the USDA and US EPA approaches to the regulation of GM plants. It was felt that the proposals by both agencies failed to take into consideration the current scientific consensus on the potential risks associated with agricultural biotechnology. Overall, the report was supportive of the science-based policy of regulation, based on the characteristics of a food product and not on the means by which it was created.

As a part of their role, each federal regulatory agency involved in assessing the human health and environmental safety of GM crops attempts to ensure that only the most current scientific data, information, and methods are referenced. The US FDA consults a Food Advisory Committee, the US EPA refers questions to Scientific Advisory Panels, and the USDA has recently established its own advisory panel. Aside from periodic reviews by credible independent research organizations such as NAS, the individual policies of each agency have been subject to regular review for harmonization with approaches recommended by international standard-setting bodies, such as OECD, WHO and FAO.

It is expected that many of the future food products of rDNA technology in agriculture may not be equivalent in composition to their conventional counterparts. However, the existing decision-tree approach to regulation anticipates these types of improved food products, and provides for the case-by-case assessment of their safety. Field testing and pre-market review for food safety provide the required level of assurance that the foods derived from the application of rDNA technology in agriculture are at least as safe as existing foods, and are consistent with all existing standards of food safety.

All new varieties of vegetables, fruit, corn and soybeans developed using rDNA technology have been subject to considerable scientific and regulatory review by the US FDA, the USDA and/or the US EPA. As such, the developers of new crop varieties have had to take into consideration not only the interests of farmers, food processors and consumers, but also the expectations of regulatory authorities.

Both regulators and consumers have had a number of opportunities to influence the acceptance of a new crop variety, and continue to have significant influence beyond final regulatory approval and commercial introduction. It is expected that the issues of human health and environmental safety will continue to be of major concern in the future as a greater number of GM food products reach later stages of commercial development.

A recent trend has been observed towards the commercialization of products that offer more transparent benefits and value to consumers. The new products of rDNA technology will be expected to further challenge the flexibility that has been exhibited by the US regulatory process. As a result, both industry and regulatory authorities will need to be responsive to uphold the human health and safety record achieved for GM product introductions. The vast majority of the new products of rDNA technology are not expected to be commercially available until after 2005.

The crops and foods improved through rDNA technology have been developed with much more precision, and their human health and environmental safety has been assessed in more depth and detail than any other crops and foods developed using more conventional means. They are routinely evaluated for any potential impact on both the environment and food safety. Despite years of experience and research, no verifiable data implicating rDNA technology as a food safety or environmental hazard have presented themselves.

The research and development of GM crops have been performed under extremely controlled conditions and tight regulatory oversight from research and development within the labouratory through field trials and commercial planting for food-use. Based on evaluations conducted to date, no evidence exists to suggest that GM foods currently sold on the market pose any significant human health concerns or that they are in any way less safe than foods derived from crops developed through more conventional means.

Although the foods developed using rDNA technology have been subject to rigorous scientific review for safety, the US regulatory system has allowed for considerable advancement in the application of the technology, and significant product innovation. To continue to be successful, it is recognized that the US FDA, USDA and US EPA will need to cooperate: the approval process must be clear, timely and transparent; emphasis must be on science-based/risk-based regulatory assessments and decisions; the regulatory process must be flexible to adapt to new information; and, the agencies must be staffed by full-time, highly qualified scientists in many diverse fields.

Over the past 25 years, the approach to regulating the products of agricultural biotechnology has evolved in response to the experience gained during the research and development of applications for rDNA technology. In the future, as the number of products developed using rDNA technology continues to increase, additional research will be needed to better evaluate any potential effects on human health and the environment, and to further refine the scientific basis of making regulatory decisions.

Within the United States, the regulatory approach to evaluating the safety of products developed using rDNA technology continues to evolve as new applications continue to emerge and our understanding of the potential risks improves. To date, existing regulations have proven adequate to assure the safety of all new products developed using rDNA technology, and additional innovation has not been discouraged. Accordingly, the National Research Council has recommended that any new rules for the regulation of GM plants be sufficiently flexible enough to reflect improvements in scientific understanding.

In the future, as rDNA technology finds broader application, and the new prospects for this technology become commercial realities, the regulatory approaches to evaluating the safety of GM food crops are expected to evolve accordingly.

5

Genetic Modification in Agriculture Sector through Biotechnology

GENETICALLY MODIFIED ORGANISMS

Genetically modified organisms (GMOs) are a fact of modern agriculture, and are here to stay. GMOs are also a fact of public preoccupation and opinion, which politicians must take into account. FAO recognizes the great potential and the complications of these new technologies. We need to move carefully, with a full understanding of all the factors involved. In particular, we need to assess GMOs is terms of their impact on food security, poverty, biosafety, and the sustainability of agriculture. Will GMOs increase the amount of food in the world, and make more food accessible to the hungry?

Clearly, GMOs should be seen not in isolation as technical achievements. Hence, we will discuss not the specifics of GMO technology, but the context in which they are developed and deployed, and about how public opinion and government policy on GMOs are formed.

The public in many countries distrusts GMOs. They are often seen in the context of globalization and of privatization and even as "antidemocratic" or "meddling with evolution".

There are as yet few perceived advantages for the public, because GMO applications to date have concentrated on reducing costs for producers without direct consumer benefits.

In particular, it has been a tactical error of the industry to concentrate on pesticide-resistance as one of the earliest applications, as this has stimulated environmental concerns. The public often confuses the industry with the science. And consumers worry about risk, not about scientific freedom. Scientists in both the private and public sectors clearly see genetic modification as a major new set of tools. They are also participants and spectators in a major shift of research from the public to the private sector, which will undoubtedly influence the future direction of research and research investment. As shareholders in the GMO debate, scientists must recognize that there is also a substantial public distrust of science.

SOCIETY OF TOXICOLOGY

The Society of Toxicology (SOT) is committed to protecting and enhancing human, animal, and environmental health through the sound application of the fundamental principles of the science of toxicology. It is with this goal in mind that the SOT defines here its current consensus position on the safety of foods produced through biotechnology. In this context, biotechnology is taken to mean those processes whereby genes that are not endogenous to the organism (transgenes) are transferred to microorganisms, plants, or animals employed in food production, or where the expression of existing genes is permanently modified, using the techniques of genetic engineering. We intentionally avoid using the term genetically modified organisms (GMOs) or foods in this context, since conventional techniques of plant and animal breeding, which are not considered here, also involve genetic modification. The extent of the genetic changes resulting from such conventional breeding techniques, which is generally undefined, far exceeds that typically produced by transgenic methods. Consequently, it is important to recognize that it is the product, and not the process of modification, that is the focus of concern regarding the human or environmental safety of biotechnology-derived (BD) foods.

The principal responsibilities of toxicologists are to define and characterize the potential for natural and manufactured materials to cause adverse health effects and to assess, as accurately as possible, the plausibility and level of risk for human or animal health or for environmental damage under a defined set of circumstances. It is not the task of the Society of Toxicology to determine the overall value of a product or process by balancing health or environmental risks with potential benefits, or to choose between different strategies to manage risk, although toxicological considerations are important in both processes. Our purpose here is rather to identify and consider the primary toxicological issues associated with BD foods. Major areas of concern in the development and application of such foods in agriculture relate to the possibility of deleterious effects on both human health and the environment. We do not consider here some aspects of the possible environmental impact of GM organisms such as gene transfer to nonengineered plants.

GENETIC MODIFICATION IN AGRICULTURE

Genetic modification, otherwise referred to as recombinant DNA (rDNA) technology or gene-splicing, has proven to be a more precise, predictable and better understood method for the manipulation of genetic material than previously attained through conventional plant breeding. To date, agricultural applications of the technology have involved the insertion of genes for desirable agronomic traits (*e.g.* herbicide tolerance, insect resistance) into a variety of crop plants, and from a variety of biological sources. Examples include soybeans modified with gene sequences from a *Streptomyces* species encoding enzymes that confer herbicide tolerance, and corn plants modified to express the insecticidal protein of an indigenous soil microorganism, *Bacillus thuringiensis* (Bt). A growing body of evidence suggests that the technology may be used to make enhancements to not only the agronomic properties, but the food, nutritional, industrial and medicinal attributes of genetically modified (GM) crops.

Regulatory supervision of rDNA technology and its products has been in place for a longer period of time in the United States than in most other parts of the world. The methods and approaches established to evaluate the safety of products developed using rDNA technology continue to evolve in response to the increasing availability of new scientific information. As our understanding of the potential applications of the technology is broadened, the safety of products developed using rDNA technology and the potential effects of introduced gene sequences on human health or the environment will be more closely scrutinized. In fact, much of the knowledge acquired during the commercialization of the products of rDNA technology in agriculture is now finding application in evaluating the safety of products developed through more conventional means.

The objective of this chapter is to provide the reader with an overview of the significant events leading up to the present, science-based, regulatory framework that exists for the safety evaluation of GM food crops within the United States. An attempt has been made to discuss concerns over the sufficiency of existing regulations, as well as to highlight recent initiatives taken by federal regulatory agencies to address them. Through better communication of how the regulatory process functions within the United States, it is anticipated that current and future applications of rDNA technology in agriculture will be met with a greater level of understanding and acceptance.

HISTORICAL PERSPECTIVE

rDNA technology was first developed in the 1970s. The initial response of the scientific community, including members of the National Academy of Science (NAS), to the prospects of rDNA technology, was to postpone any further research involving the technology until the potential risks to human health and the environment could be evaluated.

Researchers attending the International Conference on Recombinant DNA Molecules in 1975, otherwise known as the Asilomar Conference, tried to establish a scientific consensus on how best to self-regulate emerging applications of the technology. The conditions and restrictions that were proposed at this conference have formed the basis by which federal guidelines and policies for rDNA technology research were drafted within the United States.

National Institutes of Health (NIH)

The National Institutes of Health (NIH) was the first federal regulatory agency to publish their interests in evaluating the safety of rDNA technology in 1976, in the form of guidelines for the conduct of research. Because of the uncertainties that existed at the time, all research into the potential applications of rDNA technology was limited to the confines of federally funded labouratories under NIH control. After continued research, and a more careful assessment and monitoring of the risks, a set of less restrictive guidelines was published in 1978.

However, the environmental release of organisms developed using rDNA technology outside the confines of controlled labouratory conditions was prohibited unless otherwise approved by the NIH director. In the early 1980s, the NIH established an rDNA Advisory Committee (RAC) to review all data and experience gained with applications of the technology under its control. Based on recommendations of the RAC, a more relaxed set of research guidelines was published by the NIH in 1983.

The NIH approved the first environmental release of an organism developed using rDNA technology (ice-minus strain of *Pseudomonas*) in 1983. In response, they were criticized for failing to prepare a statement or assessment of the environmental impact of their regulatory decision as required under the National Environmental Policy Act (NEPA). Once the legal controversy had subsided, all responsibility that the NIH had for regulating the environmental introduction of GM organisms was relinquished. Nevertheless, NIH guidelines continue to be referenced in assessing the safety of rDNA research performed within industry, federal and other state labouratories. However, it was unclear which federal regulatory agencies would be responsible for ensuring the safety of the products developed using rDNA technology.

Office of Science and Technology Policy (OSTP)

In response to a need for clarification, the Office of Science and Technology Policy (OSTP) began work on the development of a policy to establish a federal regulatory framework for evaluating the safety of products developed using rDNA technology. According to the Coordinated Framework, the products of rDNA technology should be regulated on the basis of the unique characteristics and features that they exhibit, not their method of production. The products of rDNA technology were considered to pose risks to human health and the environment similar to those posed by conventional products already regulated within the United States. As a result, no new federal regulatory agencies or regulations were required. The Coordinated Framework did not, however, rule out the possibility of the development of new guidelines, procedures, criteria or even regulations to supplement or alter the scope of existing statutes for the products of rDNA technology.

The Coordinated Framework identified three federal regulatory agencies within the United States: the US Food and Drug Administration (US FDA), the US Department of Agriculture (USDA) and the US Environmental Protection Agency (US EPA), as having primary responsibilities for evaluating the products of rDNA technology under development at that time.

In 1992, the OSTP released another document entitled, 'Exercise of Federal Oversight within the Scope of Statutory Authority: Planned Introductions of Biotechnology Products Into the Environment', outlining the proper basis by which federal regulatory agencies were expected to exercise their regulatory authority. As with conventional products, dependent upon the intended use and function, more than one federal regulatory agency may share an interest in evaluating the safety of a product developed using rDNA technology. If more than one federal regulatory agency has an interest, lead agencies are identified as being responsible for coordinating activities to limit any potential duplication of efforts. Although federal regulatory agencies worked independent of one another, it was realised that close working relationships would need to be established in order to evaluate effectively the safety of products developed using rDNA technology.

Recently, the OSTP teamed up with the White House Council on Environmental Quality (CEQ) to perform a six-month inter-agency evaluation of the federal regulatory agency responsibilities in evaluating the environmental safety of products developed using rDNA technology. A case-study approach for a variety of different classes of products developed using rDNA technology was used to evaluate the level of federal regulatory agency involvement, to identify strengths, weaknesses and areas of potential improvement. The review concluded that none of the previously approved products of rDNA technology has had any significant negative impact on the environment.

Although all the case studies were published, OSTP/CEQ failed to reach a consensus on issues relating to the relevant strengths and weaknesses of the existing regulatory structure within the time allotted for the completion of its review. A review of the case studies published provides a comprehensive interpretation of the responsibilities of each federal regulatory agency in ensuring the safety of the products developed using rDNA technology.

National Academy of Sciences (NAS)

The National Academy of Sciences (NAS), and its operating arm, the National Research Council (NRC), have served as a primary source of scientific, technological, human health and environmental policy advice during the development of regulatory approaches for the safety evaluation of products developed using rDNA technology within the United States.

In 1987, the NRC published a report concerning the potential human health and environmental hazards associated with the commercial introduction of GM organisms, entitled *Introduction of Recombinant DNA-Engineered Organisms into the Environment: Key Issues*. The risks associated with the introduction of GM organisms were considered to be essentially the same in kind as those associated with unmodified organisms.

In other words, rDNA technology did not appear to introduce any unique risks as compared to the products that had been developed using more conventional methods of genetic modification. In reaching these conclusions, the NRC performed an evaluation of the similarities and differences in the properties exhibited by products developed using a variety of different techniques. To this day, the conclusions of this report continue to be referenced by the regulators and developers of GM crop varieties worldwide.

A subsequent NRC report, entitled *Field Testing Genetically Modified Organisms: Framework for Decisions*, also reached similar conclusions, but provided additional guidance as to how regulatory decisions concerning the introduction of GM organisms should be made. The NRC recommended that regulatory decisions concerning the introduction of GM organisms should be made on a case-by-case basis. Consistent with the Coordinated Framework, the NRC did not consider the nature of the process used for the genetic modification of an organism to be a useful criterion for determining whether a product requires less or more regulatory oversight. As a result, no valid reason existed to regulate organisms genetically modified via modern techniques (*e.g.* rDNA technology) any differently from organisms genetically modified via more conventional means. Similar conclusions have been published in the reports of international standards-setting organizations.

In retrospect, within both reports, the NRC acknowledged that modern and conventional methods of genetic modification are not without risks to human health or the environment, and as a result, neither could be considered inherently more risky. As a result, regulatory decisions concerning the safety of the products of rDNA technology need only to take into consideration the specific characteristics exhibited by a particular GM organism and the environment in which it is to be introduced, and not the method by which it has been produced. The NRC has further articulated their conclusions into what is now commonly referred to as the 'Concept of Familiarity'.

Although familiarity with the characteristics of a particular organism or the environment to which it will be introduced would not necessarily mean it was safe, it can be expected to provide a sufficient amount of information to allow for a judgement to be made of the risks. For example, familiarity with a new GM plant variety could be established based on comparisons between characteristics of the parent line or other crop species exhibiting similar traits, as well as through the results of actual field tests involving the GM plant. These principles were further elabourated upon by the Organisation for Economic Co-operation and Development (OECD), and as a result, have been referenced in the development of regulatory policies for evaluating the safety of GM crops on a global basis.

More recently, a committee established by the NRC published a report entitled *Genetically Modified Pest-Protected Plants: Science and Regulation,* based on a review of all scientific and regulatory data collected during the regulatory approval process for GM crops within the United States. The primary objective was to assess independently the effectiveness of existing and proposed regulations for the safety evaluation of GM crops expressing plant pesticides (*i.e.* plant-incorporated protectants). No new evidence was identified to suggest plants expressing plant pesticides posed any greater risk to human health or the environment as a result of their genetic modification. In fact, the NRC concluded, 'with careful planning and appropriate regulatory oversight, commercial cultivation of GM plants is not expected to pose higher risks and may pose less risk than other commonly used chemical and biological pest-management techniques'.

However, the NRC report included requests for federal regulatory agencies to further strengthen the current regulatory approval process through better coordination and communication between agencies, on-going investment in the research and monitoring of potential human health (*e.g.* allergenicity) and environmental impacts (*e.g.* insect resistance), and by providing greater access to information evaluated in support of regulatory decisions.

PRODUCTS OF AGRICULTURAL BIOTECHNOLOGY

The regulatory approach to the safety evaluation of plants developed using rDNA technology has evolved in the best interests of research scientists, industry and the general public. The agricultural products of rDNA technology, such as GM foods and crops, may require approvals from up to three regulatory agencies; the US FDA, the USDA and the US EPA, depending upon the characteristics exhibited by the GM plant, its proposed use and introduced traits. The same standards of safety are applied to all products regardless of the technology used in their development.

The US FDA is responsible for ensuring the human safety of all new foods and food components, including products developed using rDNA technology, under the Federal Food, Drug, and Cosmetic Act (FFDCA). The USDA evaluates the potential of a GM plant to become a plant pest following its environmental introduction under the Federal Plant Pest Act (FPPA). The US EPA evaluates pesticides, including plant systems modified to express pesticides (*e.g.* insect-protected or virus resistance), under the Federal Insecticide, Fungicide, and Rodenticide Act (FIFRA). As a result, the expression of an insecticidal protein in a food crop would undergo review by the USDA, US EPA and US FDA; a GM food crop exhibiting a modified oil content would be evaluated by the USDA and US FDA; and a non-food horticultural plant developed using rDNA technology for any other purpose (*e.g.* flower colour) would be subject to review by the USDA alone.

In most instances, obtaining all necessary approvals for the commercialization of an agricultural crop developed using rDNA technology takes a decade or more. However, the exact amount of time required will depend on the need to confirm performance, to evaluate characteristics of the food, environmental effects, and to produce the required amount of seed before the product can be distributed and commercially grown by farmers. Up to five years of field trials (5-10 generations of plants) are required for the developer of a new plant variety to collect sufficient data to meet the reporting requirements of the USDA. An additional five months to two years may be required for the US FDA, USDA and/or US EPA to complete all necessary product consultations, reviews and approvals.

Approval for the first commercial planting of a GM food crop was not issued until 1995. Since 1995, more than 40 new agricultural crops developed using rDNA technology have received approval for commercial planting within the United States. In 1999, approximately 72% of the total 39.9 million hectares (more than 98 million acres) of GM crops grown worldwide were planted in the United States. Herbicide-tolerant soybeans (54%), Bt corn (19%) and herbicide tolerant canola (9%) accounted for approximately 82% of the GM plants cultivated. With an increasing number of agricultural biotechnology products reaching later stages of commercial development, it is anticipated that the overall area planted with GM crops will continue to rise.

The regulatory approach to evaluating the human health and environmental safety of GM crops within the United States is best described as a science-based, case-by-case assessment of hazards and risks. This approach has provided the flexibility required to reduce the regulatory burden placed on products that have been determined to be of low risk or concern. All agencies involved in the regulation of plants developed using rDNA continue to implement and develop policies based on recommendations made within the Coordinated Framework. In the future, the US FDA, USDA and the US EPA will be dedicating additional resources towards communicating how GM food and food components are regulated within the United States, and how these regulations function to be protective of both human health and the environment.

FOOD AND FOOD COMPONENTS

The US FDA is responsible for ensuring the safety and wholesomeness of all food and food components, including the products of rDNA technology, under the FFDCA. The US FDA has the authority for the immediate removal of any product from the market that poses potential risk to public health or that is being sold without all necessary regulatory approvals. As a result, a legal burden is placed on developers and food manufacturers to ensure the commodities utilized and foods available to consumers are safe and in compliance with all legal requirements of the FFDCA. In order to understand the regulatory approach followed by the US FDA in the safety evaluation of GM crops, it is useful to consider food and food safety from a historical context.

People had been consuming foods derived from agricultural crops for many years prior to the existence of any food laws or regulations within the United States. Based on this experience, agricultural crops have been accepted as being safe for consumption as food, without additional testing to demonstrate their safety. As long as the new crop variety has exhibited similar agronomic properties, and an appropriate taste and appearance, it has been considered safe to consume. As a result, most foods consumed today, in particular whole foods (*i.e.* fruits, grains and vegetables) and conventional foods, have not been subject to any kind of premarket review or approval by the US FDA. Nevertheless, food scientists have a good understanding that many of the commonly consumed agricultural crops contain natural toxicants (*e.g.* tomatine in tomatoes, solanine in potatoes, cucurbiticin in cucumber, psoralens in celery, etc.). As a result, new plant varieties may be subject to routine chemical analyses to ensure that none of these substances is present at potentially harmful levels. This type of general approach has been used in assessing the safety of thousands of new plant varieties that have been developed over a number of decades of crop breeding without compromising the safety of whole foods.

Role

Consistent with recommendations within the Coordinated Framework, the US FDA considered existing provisions of the FFDCA to be sufficient for the regulation of foods and food components developed using rDNA technology. It was concluded that the scientific and regulatory issues posed by the products of rDNA technology were not significantly different from those posed by conventional products. As a result, GM foods and food components have been subject to the same standards of safety as already exist for the regulation of other foods and food components under the FFDCA. In order to better communicate interpretations of existing provisions of the FFDCA as they relate to the safety evaluation of foods derived from new plant varieties, including the products of rDNA technology, the US FDA released a policy statement in 1992 entitled 'Statement of policy: foods derived from new plant varieties'.

The US FDA has considered the use of genetic modification (*i.e.* rDNA technology) in the development of new plant varieties to represent a continuum of conventional plant breeding practices (*e.g.* mutagenesis, hybridization, protoplast fusion, etc.), and as a result, the safety evaluation of all new plant varieties, not just those developed using rDNA technology, have been evaluated based on an objective analysis of the characteristics of a food or its components, and not on its method of production.

Definition and Scope of Bioengineered Foods

The recently proposed rule of the US FDA concerns 'bioengineered foods' which have been defined as 'foods derived from plant varieties that are developed using *in vitro* manipulations of DNA (generally referred to as rDNA technology)'. As a result, the proposed rule has a much narrower focus than the 1992 US FDA Statement of Policy.

The US FDA has explained the need for a change in emphasis based on their expectations that many of the new plant varieties exhibit a greater potential to 'contain substances that are significantly different from, or that are present in food at a significantly different level than before'. As a result, the substances present in foods and food components derived from new plants developed using rDNA technology are less likely to be considered GRAS, and as a result, will require pre-market approval from the US FDA.

Safety and Nutritional Evaluation

The primary objective of the safety and nutritional evaluation is to demonstrate that the food derived from a new plant variety is as safe or nutritious as foods already consumed as a part of the diet. For new plant varieties, including those developed using rDNA technology, a science-based approach is used to focus the evaluation on the demonstrated characteristics of the food or food component. The evaluation of a GM food or food component typically involves reviewing information or data on any newly introduced substances, the known levels of toxicants, as well as the nutritional composition of the plant following modification. Substances that raise safety concerns (*e.g.* toxicants, allergens) would be subject to more extensive evaluation, since both intended and unintended changes may affect the levels of toxicants and nutrients in a food following the modification.

Guidance for performing a safety and nutritional evaluation was provided in the 1992 US FDA Statement of Policy, in a series of flow charts and text that cover:

- The crop that has been modified;
- Source(s) of the introduced genetic material;
- New substances intentionally added to the food as a result of the genetic modification (*e.g.* proteins, but also fatty acids, and carbohydrates).

Documentation required to support the evaluation typically includes: the purpose or intended technical effect of the modification on the plant, together with a description of the various applications or uses, a molecular characterization of the modification including the identities, sources and functions of introduced genetic material; information on the expressed protein products encoded by introduced genes; information relating to the known or suspected allergenicity and toxicity of any expressed gene products; for foods known to cause allergy, information on whether the endogenous allergens have been altered by the genetic modification; information on the compositional and nutritional characteristics of the foods, including anti-nutrients; and in some instances, comparative results of feeding studies involving the foods derived from plants modified using rDNA technology and the non-modified counterpart.

In performing its evaluation, the US FDA is particularly interested in the identification of inherent toxicants, known or potential allergens, assessing the concentration and bioavailability of essential nutrients, the safety and nutritional value of any newly introduced proteins, and the identity, composition, and nutritional value of modified carbohydrates, fats and oils.

If additional questions of safety remain following this evaluation, further toxicological studies may need to be performed. It is recognized that absolute assurance of the safety of any food does not exist. As a result, the goal of the safety evaluation is to establish a reasonable certainty of no harm under anticipated conditions of consumption. With this in mind, experience with the existing food supply has provided the basis for evaluating the safety of new food or food components. Both the Food Advisory Committee and Committee for Veterinary Medicine have been extensively involved in the development of approaches for the safety and nutritional evaluation of foods and food components derived from new plant varieties, including those developed using rDNA technology.

GENETIC MATERIAL IN PRODUCT CHARACTERIZATION

Product characterization takes into consideration information relating to the modified food crop, the introduced genetic material and its expression product, and acceptable levels of inherent plant toxicants and nutrients. All characteristics of the gene insert must be known, including the source(s), size, number of insertion sites, promoter regions, and marker sequences. It must be established that the transferred genetic material does not come from a pathogenic source, a known source of allergens, or a known toxicant-producing source. The introduced genetic material should be well characterized to ensure that the introduced gene sequences do not encode harmful substances and are stably inserted within the plant genome to minimize any potential opportunity for undesired genetic rearrangement.

Analytical data are required to evaluate the nutritional composition, the levels of any known toxicants, anti-nutritional and allergenic substances, and the safety-in-use of antibiotic resistance marker genes. Any new substances introduced into crops through rDNA technology (*e.g.* proteins, fatty acids, carbohydrates) will be subject to pre-market review as food additives by the US FDA, unless substantially similar to substances already safely consumed as a part of foods, or that are considered GRAS. To date, substances that have been added to foods through rDNA technology have been previously consumed or have been determined to be substantially similar to substances already consumed as a part of the diet. As such, introduced substances have been considered exempt from the requirement for pre-market approval as food additives with the US FDA.

A more rigorous safety evaluation of a GM crop is warranted if the introduced gene sequence(s) has not been fully characterized, the nutritional composition has been significantly altered, antibiotic resistance marker genes have been used during its development, or if an allergenic protein or toxicant has been detected at levels higher than what is typically observed in edible varieties of the same crop species. In any event, determinations as to the safety of substances that have been introduced into new plant varieties through rDNA technology are made on a case-by-case basis.

Although an evaluation of the introduced gene sequences and expression product(s) provides assurance as to their safety, further studies may be required to predict whether unexpected effects may result following their interaction with other genes within the plant. In addition, the product characterization of a GM plant involves assessing sequence homology to known toxicants and allergens, thermal and digestive stability, and if required, the results of both *in vitro* and *in vivo* assays to demonstrate lack of toxicity.

COMPOSITIONAL ANALYSIS

The results of field trials performed over several years serve to characterize the phenotypic and agronomic characteristics exhibited by the plant (*e.g.* height, colour, leaf orientation, susceptibility to disease, root strength, vigour, fruit or grain size, yield, etc.), as well as to provide the materials required for the compositional analysis. Any anomalies in the phenotypic or agronomic characteristics exhibited by a plant may result in a requirement for additional information. Protein, fat, fibre, starch, amino acid, fatty acid, ash and sugar levels are determined, as well as the levels of anti-nutrients, natural toxicants or known allergens. Studies of the nutritional composition are performed to determine whether the levels of any key nutrients, vitamins or minerals have been altered as a result of the genetic modification.

Based on the results of these studies, a determination is made as to whether the phenotypic and agronomic characteristics of a GM crop or the concentrations of inherent constituents fall within ranges typical of its conventional counterpart. If inserting a new gene causes no change in any of the assessed parameters, the US FDA can conclude with reasonable assurance that the GM crop is as safe as the conventional crop. If the levels of essential nutrients or inherent toxicants are found to be significantly different in the GM crop, the US FDA may recommend additional action prior to commercialization, such as obtaining food additive status, or the use of specific labels to alert consumers of an altered nutritional content, etc.

ALLERGENICITY

In consultation with scientific experts in the areas of food safety, food allergy, immunology, biotechnology and diagnostics, the US FDA published guidelines for assessing the allergenicity of GM foods or food components in 1994.

The approach to assessment is multi-faceted, incorporating data regarding the origin of the genetic material, and the biochemical, immunological and physicochemical properties of the expressed protein. The overall assessment is reliant upon the fact that all known food allergens are proteins and, notwithstanding the number of shared properties between allergenic and non-allergenic food proteins, food allergens tend to exhibit a number of similar characteristics. In general, food allergens share a number of common properties: they have a molecular weight of over 10 000 Da; they represent more than 1% of the total protein content of the food; they demonstrate resistance to heat, acid treatment, proteolysis and digestion; and they are recognized by IgE.

For gene sequences derived from known allergenic sources (*e.g.* peanuts), the developers of GM plants are expected to demonstrate that allergenic proteins have not been introduced into the food. For assessment purposes, it is assumed that any genetic material derived from a known allergenic source will encode for an allergen. To demonstrate otherwise, the amino acid sequence of an expressed protein must be compared with that of known allergens using protein sequence databases. Furthermore, *in vitro* and/or *in vivo* immunologic analyses using the sera of allergic patients sensitive to the source of the genetic material may need to be performed to determine whether or not a potentially allergenic protein is being expressed in the GM food.

Some GM foods may be modified to express genes from a source that is not known to be allergenic when consumed. Under these circumstances, the US FDA follows a similar decision tree-based approach to determining the allergenic potential of the expression product. In assessing these proteins, should any amino acid sequence exhibit homology with a known allergen, the expressed protein would then be evaluated in immunologic tests using the sera of patients known to be allergic to the identified homologous protein. Regardless of the origin of the genetic material, physicochemical studies are performed *in vitro* to provide information concerning the expected stability of the expressed protein. All known food allergens tend to be resistant to digestive degradation, as demonstrated in simulated gastric fluid models, or to decomposition under conditions of food processing.

In making a determination regarding a GM food, it is the totality of the biochemical, immunological and physicochemical properties of the introduced protein that provides guidance as to the allergenic potential of such a protein being expressed in food. The level of protein expressed, produced and consumed as a part of the diet is also a primary indicator of the allergenic potential, since nearly all food allergens are known to be major proteins in their respective foods. If the results of any of these studies suggest an allergenic potential, the US FDA may recommend further scientific evaluation, require special labelling to alert sensitive consumers, or alternatively caution the developer about proceeding with the development of a particular GM food.

Most recently, a joint FAO/WHO Expert Consultation on Allergenicity of Foods Derived from Biotechnology recommended a revised decision tree. This decision-tree strategy was modified from the previous version FAO/WHO to include a revised definition of sequence homology for gene product comparison; a greater emphasis on serum testing, even with gene products without homology to known allergens and not derived from an allergenic source; and animal models to assess potential allergenicity, despite acknowledgement by the Expert Consultation that these models are currently under development and at present, not predictive of food allergies in humans.

More recently, the *ad-hoc* Open-Ended Working Group on Allergenicity established by the *ad-hoc* Intergovernmental Codex Task Force on Foods Derived from Biotechnology, considered the FAO/WHO strategy in drafting an approach to assessing the potential allergenicity of foods derived from rDNA plants.

The Codex Working Group recognized the absence of a definitive predicitve test for allergenicity in humans to a newly expressed protein and recommended an integrated, stepwise assessment strategy. The strategy recommended by the Working Group, and accepted by the Codex Task Force for inclusion in the draft forwarded for final adoption by the Codex Alimenarius Commission, is consistent with that of FAO/WHO. However, the Working Group suggested that some of the modifications included in FAO/WHO could contribute to the overall weight of evidence of any conclusion of potential allergenicity (*e.g.*, allergen specific serum depositories, animal models), pending development and validation.

ANTIBIOTIC RESISTANCE

The use of antibiotic-resistance genes as selectable markers has been common practice in the development of new plant varieties using rDNA technology. Concerns relate to the potential transfer of antibiotic-resistance genes from GM plants to pathogens in the environment or to the gut of humans consuming foods or food components. Issues relating to the use of antibiotic resistance genes were identified in the 1992 US FDA Statement of Policy Statement as well as discussed in additional guidance entitled *Guidance for Industry: Use of Antibiotic Resistance Marker Genes in Transgenic Plants*. The guidance provided within these documents was established in consultation with experts in the fields of microbiology, medicine, food safety, bacterial and mycotic diseases, and includes suggestions with respect to the continued safe use of antibiotic-resistance marker genes by the developers of new plant varieties.

The use of marker genes that encode resistance to clinically important antibiotics has raised questions as to whether their presence in food could reduce the effectiveness of oral doses of the antibiotic or whether the gene present in the DNA could be transferred to pathogenic microbes, rendering them resistant to treatment with the antibiotic. The risk of transfer of antibiotic-resistance genes from plants to microorganisms considered to be pathogenic to humans, however, is considered to be minimal if not insignificant. Furthermore, the potential risks are becoming less of a concern as more developers are beginning to research the use of alternative technologies (*e.g.* non-resistance-based markers) in plant breeding. The conclusions with respect to the safe use of antibiotic-resistance marker genes are consistent with the findings of other national and international food safety organizations.

Consultation and Filing Process

The submission of a Pre-market Biotechnology Notification (PBN) has recently been proposed as a mandatory requirement for the commercialization of bioengineered foods and food ingredients within the United States. Minimal differences exist between the information to be submitted as a part of a PBN and that previously presented in voluntary consultations with the US FDA. As proposed, developers are still being encouraged to consult with the US FDA as early and as often as necessary in the development of a bioengineered food, such that any potential scientific or regulatory concerns can be identified and addressed prior to the submission of the PBN.

Guidelines for performing consultations with the US FDA were released in a publication entitled *Guidance On Consultation Procedures: Foods Derived from New Plant Varieties* in 1997. The publication recommended an approach for developers to proceed by submitting a request for consultation, outlined the internal process by which all requests would be handled, and included additional guidance as to the type of safety and nutritional information to be presented during consultation with the US FDA.

Once sufficient safety and nutritional information has accumulated to demonstrate that a product is safe and in compliance with the FFDCA, developers typically schedule a consultation to present their scientific findings and conclusions to the US FDA. Consultations prior to notification not only serve to keep the US FDA informed of advances made in the application of rDNA technology in food production, but also keep the developers of bioengineered foods aware of emerging safety, nutritional or regulatory concerns of the US FDA.

Consultations are considered complete when all safety and regulatory concerns between the US FDA and the developer have been resolved. The US FDA proposes to perform an initial evaluation of a PBN within 15 days of receipt to determine completeness, at which point, if considered complete, the PBN will be filed, and a response can be expected within 120 days. The US FDA does not issue a product approval *per se*, but informs the developer by letter that:

- The evaluation period has been extended;
- The notice is not complete and why;
- It has no further questions 'at this time' based on the information that has been presented.

Labelling

The FFDCA defines what information must be disclosed to consumers on a food label, such as the common or usual name, and other limitations concerning the representations or claims that can be made or suggested about a food product. All foods must be labelled truthfully and not be misleading to consumers. Taking this into consideration, the FFDCA does not stipulate the disclosure of information on the basis of consumer desire to know. Labelling may be considered misleading if it fails to reveal material facts in light of representations that are made with respect to a product.

The labelling of foods derived from new plant varieties, including plants developed using rDNA technology, was originally addressed in the 1992 US FDA Statement of Policy, and most recently discussed in Draft Guidance for Industry for the voluntary labelling of bioengineered foods.

To date, the US FDA is not aware of any information that would distinguish foods developed using rDNA technology (*e.g.* bioengineered foods) as a class from foods developed through other methods of conventional plant breeding, and as such, have not considered the method of development a material fact requiring disclosure on product labels. Nevertheless, after extensive consultation, including thousands of written comments and a series of public meetings, the US FDA has observed 'a general agreement that providing more information to consumers about bioengineered foods would be useful'.

Requirements for Labelling

Special labelling is required if the composition of the bioengineered food differs significantly from its conventional counterpart. For example, for a food that has been genetically modified to contain a new major sweetener, a new common or usual name or other labelling may be required. Similarly, if a GM food contains an allergen that consumers would not expect to be present in that food, special labelling may be necessary to alert sensitive consumers. If a protein commonly associated with an allergic reaction (*e.g.* peanut protein) is transferred to another food through genetic modification, the US FDA would evaluate whether labelling would provide sufficient consumer protection. If labelling would not be considered to provide a sufficient level of protection, the US FDA would take appropriate steps to ensure the GM food would not be marketed. Therefore, current policy requires a GM food to be labelled when the resulting product poses a safety issue or is substantially different from its conventional counterpart, and as a result, could be considered to pose a misrepresentation to consumers.

The 1992 US FDA Statement of Policy and the recent Draft Guidelines do not consider the use of rDNA technology in the development of food products to be a material fact requiring specific disclosure on the label. Rather it is a method of development, similar to other methods of plant breeding, which have not required disclosure on the label. Bioengineered foods cannot be distinguished compositionally from foods modified through more conventional methods and thus do not require specific disclosure through labelling. The Draft Guidelines reaffirm the US FDA position that bioengineered foods do not require special labelling.

Voluntary Labelling

To provide guiding principles for voluntary labelling, in recognition of the desire of certain manufacturers to label foods as produced either with or without bioengineering, the US FDA published a 'Draft Guidance for Industry: voluntary labelling indicating whether the foods have or have not been developed using bioengineering'. Emphasizing that the use of rDNA technology 'is not a material fact', the US FDA recognizes that some consumers want disclosure of bioengineered content and that some manufacturers wish to provide it. In response, Draft Guidance was issued with suggestions concerning the use of labelling statements that are not considered misleading.

PRESENCE OR USE OF RDNA

The US FDA provided several examples of how disclosure of bioengineering can be accomplished, be informative and not be misleading.

- *Example* 1 'Genetically engineered' or 'This product contains corn meal that was produced using biotechnology'. These disclosures reveal the minimum amount of optional information about bioengineering.
- *Example* 2 'This product contains high oleic acid soybean oil from soybeans developed using biotechnology to decrease the amount of saturated fat.' This statement explains the proper and required name of this type of soybean oil, since it differs from what would

be considered as soybean oil. The optional comments about biotechnology and decreasing saturated fat provide information that could be seen as benefits of the product.

- *Example* 3 'These tomatoes were genetically engineered to improve texture.' This example was one included to illustrate how it could be misleading to consumers if they cannot discern a difference in texture, but would not be misleading if they can tell a difference. If the former, and the new texture is to facilitate processing, then this intention should be made clear, *i.e.* 'to improve texture for processing'.
- *Example* 4 'Some of our growers plant tomato seeds that were developed through biotechnology to increase crop yield.' This is another example of optional information that would explain an indirect, agricultural benefit.

ABSENCE OR NOT BIOENGINEERED

The US FDA provides an important commentary in the Draft Guidance about 'genetically modified organisms (GMO)' versus 'bioengineered'. It would be technically inaccurate to use the phrase 'not genetically modified' or 'GMO-free' to mean that bioengineering was not used. This is because most conventional foods have been genetically modified over the years by traditional crop breeding practices. Examples of acceptable voluntary statements would include:

- We do not use ingredients that were produced by biotechnology.'
- This oil is made from soybeans that were not genetically engineered.'
- Our tomato growers do not plant seeds developed using biotechnology.'

Another important point is made about a term such as 'GMO-free' that is misleading for two reasons:

- Many foods do not contain organisms anyway and therefore should not be labelled as 'organism-free';
- Free' implies complete absence or 'zero' amount of bioengineered material.

In a practical sense, it is impossible to demonstrate analytically the complete absence of anything. Therefore, in reality, a threshold for possible adventitious presence of a low level of bioengineered material may be necessary and has been the subject of much debate. The Agency also provides additional guidance about misleading statements that:

- Could be interpreted to suggest that the absence of bioengineering would make the food superior to the bioengineered alternative;
- Claim the absence of one bioengineered ingredient when the food contains another ingredient that is bioengineered;

- Claim that a food is not bioengineered when in fact this type of food (*e.g.* green beans) has never been modified through rDNA technology.

TYPES OF TOXICOLOGICAL HAZARDS TO CONSUMERS AND PRODUCERS ASSOCIATED WITH BD FOODS

Current techniques of developing organisms used in the production of BD foods typically involve the transfer to the host of the desired gene or genes in combination with a promoter and a gene for a selectable marker trait that allows the efficient isolation of cells or organisms that have been transformed from those that have not. Common selectable markers in plants have included resistance to antibiotics (kanamycin/neomycin or ampicillin) or herbicides.

Several key issues have been raised with respect to the potential toxicity associated with BD foods, including the inherent toxicity of the transgenes and their products, and unintended (pleiotropic or mutagenic) effects resulting from the insertion of the new genetic material into the host genome. Unintended effects of gene insertion might include an over-expression by the host of inherently toxic or pharmacologically active substances, silencing of normal host genes, or alterations in host metabolic pathways. It is important to recognize that, with the exception of the introduction of marker genes, the process of genetic engineering does not, in itself, create new types of risk.

FUTURE CHALLENGES IN THE ASSESSMENT OF THE SAFETY OF BD FOODS

Current safety assessment methodologies are focused primarily on the evaluation of the toxicity of single chemicals. Food is a complex mixture of many chemicals. Using animal models, the evaluation of most aspects of the safety of single components of the diet, such as a Bt toxin, is possible using widely accepted protocols. Future projects may involve more complicated manipulations of plant chemistry. In this case, safety testing will be more challenging. Whole foods cannot be tested with the high dose strategy currently used for single chemicals to increase the sensitivity in detecting toxic endpoints. Also, the question of potential deleterious interactions between new or enhanced levels of known toxic agents in BD foods will undoubtedly be raised. The safety testing of multiple combinations of chemicals remains a difficult proposition for toxicologists. In view of these challenges, there is a clear need for the development of effective protocols to allow the assessment of the safety of whole foods. The responsibility of toxicologists is to assess whether foods derived through biotechnology are at least as safe as their conventional counterparts and to ascertain that any levels of additional risk are clearly defined. In achieving this goal, it is important to recognize that it is the food product itself, rather than the process through which it is made, that should be the focus of attention. In assessing safety, the use of the substantial equivalency concept provides guidance as to the nature of any new hazards.

Scientific analysis indicates that the process of BD food production is unlikely to lead to hazards of a different nature from those already familiar to toxicologists. The safety of current BD foods, compared with their conventional counterparts, can be assessed with reasonable certainty using established and accepted methods of analytical, nutritional, and toxicological research.

A significant limitation may occur in the future if transgenic technology results in more substantial and complex changes in a foodstuff. Methods have not yet been developed by which whole foods (as compared with single chemical components) can be fully evaluated for safety. Progress also needs to be made in developing definitive methods for the identification and characterization of protein allergens, and this is currently a major focus of research. Improved methods of profiling plant and microbial metabolities, proteins, and gene expression may be helpful in detecting unexpected changes in BD organisms and in establishing substantial equivalence. The level of safety of current BD foods to consumers appears to be equivalent to that of traditional foods. Verified records of adverse health effects are absent, although the current passive reporting system would probably not detect minor or rare adverse effects, nor can it detect a moderate increase in common effects such as diarrhea. However, this is no guarantee that all future genetic modifications will have such apparently benign and predictable results. A continuing evolution of toxicological methodologies and regulatory strategies will be necessary to ensure that this level of safety is maintained.

PRODUCTION OF TOXINS

Some bacterial toxins are utilized as invasins because they act locally to promote bacterial invasion. Examples are extracellular enzymes that degrade tissue matrices or fibrin, allowing the bacteria to spread. This includes collagenase, hyaluronidase and streptokinase. Other toxins, also considered invasins, degrade membrane components, such as phospholipases and lecithinases. The pore-forming toxins that insert a pore into eucaryotic membranes are considered as invasins.

The level of risk of these gene products to consumers and those involved in food production can be and is evaluated by standard toxicological methods. The toxicology testing for the Bt endotoxins typifies this approach and has been described in detail by the U.S. EPA. The safety of most Bt toxins is assured by their easy digestibility as well as by their lack of intrinsic activity in mammalian systems. In this case, the good understanding of the mechanism of action of Bt toxins, and the selective nature of their biochemical effects on insect systems, increases the degree of certainty of the safety evaluations. However, each new transgenic product must be considered individually, based on exposure levels and its potency in causing any toxic effects, as is typical of current risk assessment paradigms for chemical agents.

BACTERIAL TOXIGENESIS

Toxigenesis, or the ability to produce toxins, is an underlying mechanism by which many bacterial pathogens produce disease. At a chemical level, there are two main types of bacterial toxins, lipopolysaccharides, which are associated with the cell wall of Gram-negative bacteria, and proteins, which are released from bacterial cells and may act at tissue sites removed from the site of bacterial growth. The cell-associated toxins are referred to as endotoxins and the extracellular diffusible toxins are referred to as exotoxins.

Endotoxins are cell-associated substances that are structural components of bacteria. Most endotoxins are located in the cell envelope. In the context of this article, endotoxin refers specifically to the lipopolysaccharide (LPS) or lipooligosaccharide (LOS) located in the outer membrane of Gram-negative bacteria. Although structural components of cells, soluble endotoxins may be released from growing bacteria or from cells that are lysed as a result of effective host defence mechanisms or by the activities of certain antibiotics. Endotoxins generally act in the vicinity of bacterial growth or presence.

Exotoxins are usually secreted by bacteria and act at a site removed from bacterial growth. However, in some cases, exotoxins are only released by lysis of the bacterial cell. Exotoxins are usually proteins, minimally polypeptides, that act enzymatically or through direct action with host cells and stimulate a variety of host responses. Most exotoxins act at tissue sites remote from the original point of bacterial invasion or growth. However, some bacterial exotoxins act at the site of pathogen colonization and may play a role in invasion.

PRODUCTION OF ALLERGENS

Allergenicity is one of the major concerns about food derived from transgenic crops. However, it is important to keep in mind that eating conventional food is not risk-free; allergies occur with many known and even new conventional foods. For example, the kiwi fruit was introduced into the U.S. and the European markets in the 1960s with no known human allergies; however, today there are people allergic to this fruit. The issues that have to be addressed regarding the potential allergenicity of BD foods are:

- Do the products of novel genes have the ability to elicit allergic reactions in individuals who are already sensitized to the same, or a structurally similar, protein?
- Will transgenic techniques alter the level of expression of existing protein allergens in the host crop plant?

- Do the products of novel genes engineered into food plants have the ability to induce *de novo* sensitization among susceptible individuals?

Considerable scientific resources are being committed to determine the most appropriate and accurate approaches for identifying and characterizing potentially allergenic proteins.

The first systematic approach to allergenicity assessment was developed by the International Life Sciences Institute (ILSI) in collaboration with the International Food Biotechnology Council and was published in 1996. The hierarchical approach described therein has been reviewed and revised by the World Health Organization (WHO) and the Food and Agriculture Organization of the United Nations (FAO) (FAO/WHO, 2001b).

The main approaches currently used in the evaluation of allergenicity are:

- *Determinations of structural similarity, sequence homology, and serological identity*. The objective is to determine whether, and to what extent, the novel protein of interest resembles other proteins that are known to cause allergy among human populations. There are essentially three generic approaches. The first is to examine the overall structural similarity between the protein of interest and known allergens. The second is to determine, using appropriate databases, whether the novel protein is similar to known allergens with respect to either overall amino acid homology, or to discrete areas of the molecule where complete sequence identity with a known allergen may indicate the presence of shared epitopes. The third approach is to determine whether specific IgE antibodies in serum drawn from sensitized subjects are able to recognize the protein of interest.
- *Assessment of proteolytic stability*. There exists a good, but incomplete, correlation between the resistance of proteins to proteolytic digestion and their allergenic potential, the theory being that relative resistance to digestion will facilitate induction of allergic responses, provided the protein possesses allergenic properties. One approach, therefore, is to characterize the susceptibility of the protein of interest to digestion by pepsin or in a simulated gastric fluid. However, this approach alone may not be sufficient to identify cross-reactive proteins with the potential to elicit allergic responses in food-or latex-sensitized individuals as in the case of oral allergy syndrome or latex-fruit syndrome. Nor are considerations of stability to digestion necessarily relevant for allergens that act through dermal or inhalation exposure and that may have significance for worker health. In these cases, other approaches such as structural homology searches and the use of animal models may be effective in identifying potential new allergens.

- *Use of animal models.* Currently there are no widely accepted or thoroughly evaluated animal models available for the identification of protein allergens. Nevertheless, progress is being made and methods based on the characterization of allergic responses or allergic reactions in rodents and other species have been described.

Although testing strategies for allergens are still evolving and no single test is fully predictive of human responses, the approaches outlined above, when used in combination, allow scientists to address questions of potential allergenicity, and these will increase in precision and certainty with time. Considerations of this type led U.S. federal agencies to deny approval of StarLink corn for human consumption because of the possibility that its Bt protein, Cry9C, may be a human allergen. This protein had been modified to slow its digestion and prolong its effect in the insect gut and this change rendered the protein less digestible in the human gut as well. After the accidental introduction of StarLink corn into the human food chain, a limited number of illnesses among consumers were reported. These were investigated by the Centres for Disease Control, who found no evidence that the corn products were responsible. However, although this study is reassuring, methodological limitations make it less than conclusive, and it cannot eliminate the possibility that some adverse effects may have occurred that were not reported. Because of this incident, StarLink corn is no longer marketed. With the exception of Cry9C, none of the engineered proteins in foods so far evaluated through the FDA consultation process has had the characteristics of an allergen.

The only documented case where a human allergen was introduced into a food component by genetic engineering occurred when attempts were made to improve the nutritional quality of soybeans using a brazil nut protein, the methionine-rich 2S albumin. Allergies to the brazil nut have been documented, and while still in precommercial development, testing of these new soybeans for allergenicity was conducted in university and industrial laboratories. It was found that serum from people allergic to Brazil nuts also reacted to the new soybean. Once this was discovered, further development of the new soybean variety was halted and it was never marketed. This work led to the identification of the major protein associated with Brazil nut allergy, which was previously unknown.

Will Insertion of the Transgene Increase the Potential Hazard from Toxins or Pharmacologically Active Substances Present in the Host?

Concern has been expressed about the randomness with which genes are inserted into the host by current genetic engineering processes. This could, and does, result in pleiotropic and insertional mutagenic effects. The former term refers to the situation where a single gene causes multiple changes in the host phenotype and the latter to the situation where the insertion of the new gene induces changes in the expression of other genes. Such changes due to random insertion might cause the silencing of genes, changes in their level of expression, or, potentially, the turning on of existing genes that were not previously being expressed. Pleiotropic effects could be manifested as unexpected new metabolic reactions arising from the activity of the inserted gene product on existing substrates or as changes in flow rates through normal metabolic pathways.

Unexpected and potentially undesirable pleiotropic or mutagenic changes in the genome of the host do occur, but these would likely be revealed by their effects on the development, growth, or fertility of the host, or by the extensive testing of its chemical composition compared with isogenic untransformed plants, which is a necessary part of any safety evaluation of transgenic crops.

In the U.S., since 1987, the USDA Animal and Plant Health Inspection Service has completed over 5000 field trials with more than 70 different transgenic plant species. The only unexpected result was a mutation in a colour gene and gene silencing through changes in the methylation status of these genes that led to unexpected colour patterns in petunia flowers. Both of these effects are also seen in conventional plant breeding. While the possibility of an undetected increase in a toxic component in a new food cannot be entirely eliminated, the current safeguards make this unlikely, and no toxicologically or nutritionally significant changes of this type are evident in the transgenic plants so far marketed for food production.

Substantial public concern about the safety of BD products was raised in 1989 when a number of cases of eosinophilia-myalgia syndrome (EMS) were reported among users of the amino acid tryptophan as a dietary supplement. By mid-1993, 37 deaths had been attributed to this outbreak. The development of the syndrome appeared among users of some batches of the supplement after a change in the manufacturing process that included the use of a new genetically modified microorganism in the fermentation. However, concomitant with this change were additional alterations in certain filtration and purification steps used previously in the manufacturing process. The exact cause of the outbreak and the nature of the toxic impurity have not been established with certainty.

Thus, it is not possible to determine whether the change in purification, the genetic engineering of the organism, or some other factor or factors were to blame. A subsequent investigation revealed that cases of EMS also occurred among consumers of tryptophan before the GM organism was introduced into the manufacturing process, although at a lower incidence. Thus, the genetic modifications might have caused an increase in the level of the agent that was responsible for tryptophan-associated EMS, but it did not create a novel toxicant. This event is troubling in that the tryptophan would be regarded as highly purified (99.6% or higher), and no adequate animal model has been found to replicate EMS, a probable autoimmune disease. This illustrates that toxicology has limits in its ability to explain and predict adverse effects in humans.

These examples indicate that careful analysis of the changes in BD organisms is necessary to ensure against unexpected alterations in the levels of toxins, allergens, and essential nutrients. This analysis will be particularly critical if, as seems likely, engineering of the synthetic pathways of secondary metabolites is undertaken in plants, *e.g.*, to increase their resistance to insects and pathogens or to produce compounds of pharmaceutical value. Such changes might create new and unanticipated secondary compounds with unknown toxic properties. New approaches to profiling changes in metabolites, proteins, and gene expression may be helpful in such cases.

Does the Possible Transfer of Antibiotic Resistance Marker Genes from the Ingested BD Food to Gut Microbes Present a Significant Human Hazard? The development of antibiotic resistance among pathogenic bacteria is a significant human health issue. However, no contribution to antibiotic resistance in gut bacteria arising from antibiotic resistance markers in BD foods has been documented. For several reasons, including the efficient destruction of the resistance gene in the human gut and the very low intrinsic rate of plant–microbe gene transfer, any contribution from this source is expected to be extremely small. Genes for resistance to kanamycin and related antibiotics already occur quite commonly in the environment, including in the flora of the human gut, which naturally contains about 1 trillion (10^{12}) kanamycin-or neomycin-resistant bacteria. Even if the occasional transfer of resistance from plant to bacterium did occur, the practical impact would be negligible. However, since any increase in antibiotic resistance is recognized as undesirable and the technology is now available to omit the use of such marker genes, future genetically modified organisms are unlikely to contain them. Thus, concerns related to their use are likely to diminish.

Will Genetic Transformation Adversely Affect the Nutritional Value of the Host? In the USA, the FDA is entrusted with assuring that the nutritional composition of BD foods is substantially equivalent to that of the nonmodified food. Studies are performed to determine whether nutrients, vitamins, and minerals in the new food occur at the same level as in the conventionally bred food sources. A typical example is the case of Roundup Ready soybeans. In this case, the protein, oil, fibre, ash, carbohydrates, and moisture content and the amino acid and fatty acid composition in seeds and toasted soybean meal were compared with conventional soybeans. Fatty acid compositions and protein or amino acid levels of soybean oil were compared and special attention was given to checking the levels of antinutrients typically found in soybeans, *e.g.*, trypsin inhibitors, lectins, and isoflavones.

One difference between the conventional and nonconventional soybeans was detected in defatted, nontoasted soybean meal, the starting material for commercially utilized soybean protein, which is not itself consumed. In this material, trypsin inhibitor levels were 11–26% higher in the transgenic soybeans. The levels of the trypsin inhibitors were similar in all lines in the seeds and in defatted, toasted soybean meal, the form used in foods. Except for this difference in trypsin inhibitor levels, all other nutritional aspects were equivalent between the transgenic line and the conventional soybean cultivars. Feeding studies demonstrated that there were no evident differences in nutritional value between the conventional and transgenic soybeans in rats, chickens, catfish, and dairy cattle. Domestic animal feeding studies with a number of other transgenic crops have similarly shown no significant adverse changes in nutritional value.

Will the Transgene Product Adversely Affect Nontarget Organisms? In addition to the general concerns addressed that relate to food safety, additional attention is needed when the gene product is pesticidal or otherwise may be toxic to nontarget organisms that consume it. The effects of each transgene product that is designed for pesticidal effects must be evaluated on a case-by-case basis against target and nontarget organisms under specific field growth conditions for each transgenic crop.

The foremost current example of this is the incorporation of Bt genes into crop plants for insect control. The toxic properties of Bt endotoxins to both target and nontarget species of many kinds are well known. They show a narrow range of toxicity limited to specific groups of insects, primarily Lepidoptera, Coleoptera, or Diptera, depending on the Bt strain. Nevertheless, Bt-producing plants have been tested broadly to determine whether any alteration in this limited spectrum of toxicity has occurred, without the discovery of any unexpected results. Exotoxins and enterotoxins, which are much more broadly toxic than the endotoxins, are also produced by some Bt strains, but these are not present in the transformed plant, because their genes are not transferred into the crop.

In plants transformed with Bt genes to control lepidopterans, toxicity to nontarget lepidopterans would be expected if exposure occurs by feeding on the transformed crop. Particular concern has been expressed over the potential toxicity of the Bt toxin in corn pollen to the Monarch butterfly after initial laboratory studies showed increased mortality in larvae fed on leaves dusted with transgenic pollen. However, most transgenic corn pollen contains much lower nonlethal levels of Bt toxins than the strain used in this study, and there is only a limited synchrony between the feeding period of the most sensitive younger larvae and the period when corn pollen is shed.

Also, corn pollen does not typically move far beyond the borders of the field, leaving significant amounts of milkweed uncontaminated in many locations. For these reasons, a detailed risk assessment concluded it is unlikely that a substantial risk to these butterflies exists in the field since only a negligible portion of the population is exposed to toxic levels of Bt. Beyond the question of the potential toxicity of Bt corn to such valued insects, it is also important to recollect that the common alternative is to spray corn with synthetic insecticides, which are not as selective as the Bt toxin. In a sweet corn field containing milkweed plants and treated with a synthetic pyrethroid for insect control, 91–100% of the monarch butterfly larvae placed on the milkweed leaves after spraying were killed. In plots where Bt sweet corn was planted and the pollen fell naturally on the milkweed leaves, larval death rates were much lower (7–20%) and indistinguishable from those in untreated non-Bt corn plots.

GENETICALLY MODIFIED ORGANISM AND GENETICALLY ENGINEERED ORGANISM

A genetically modified organism (GMO) or genetically engineered organism (GEO) is an organism whose genetic material has been altered using genetic engineering techniques. These techniques, generally known as recombinant DNA technology, use DNA molecules from different sources, which are combined into one molecule to create a new set of genes. This DNA is then transferred into an organism, giving it modified or novel genes. Transgenic organisms, a subset of GMOs, are organisms which have inserted DNA that originated in a different species. Some GMOs contain no DNA from other species and are therefore not transgenic but cisgenic.

PRODUCTION

Genetic modification involves the insertion or deletion of genes. When genes are inserted, they usually come from a different species, which is a form of horizontal gene transfer. In nature this can occur when exogenous DNA penetrates the cell membrane for any reason. To do this artificially may require attaching the genes to a virus or just physically inserting the extra DNA into the nucleus of the intended host with a very small syringe, or with very small particles fired from a gene gun. However, other methods exploit natural forms of gene transfer, such as the ability of *Agrobacterium* to transfer genetic material to plants, or the ability of lentiviruses to transfer genes to animal cells.

History

The general principle of producing a GMO is to add new genetic material into an organism's genome. This is called genetic engineering and was made possible through the discovery of DNA and the creation of the first recombinant bacteria in 1973, *i.e., E.coli* expressing a Salmonella gene. This led to concerns in the scientific community about potential risks from genetic engineering, which were thoroughly discussed at the Asilomar Conference. One of the main recommendations from this meeting was that government oversight of recombinant DNA research should be established until the technology was deemed safe. Herbert Boyer then founded the first company to use recombinant DNA technology, Genentech, and in 1978 the company announced creation of an *E. coli* strain producing the human protein insulin.

In 1986, field tests of bacteria genetically engineered to protect plants from frost damage (ice-minus bacteria) at a small biotechnology company called Advanced Genetic Sciences of Oakland, California, were repeatedly delayed by opponents of biotechnology. In the same year, a proposed field test of a microbe genetically engineered for a pest resistance protein by Monsanto Company was dropped.

Uses

GMOs have widespread applications. They are used in biological and medical research, production of pharmaceutical drugs, experimental medicine (*e.g.* gene therapy), and agriculture (*e.g.* golden rice). The term "genetically modified organism" does not always imply, but can include, targeted insertions of genes from one species into another. For example, a gene from a jellyfish, encoding a fluorescent protein called GFP, can be physically linked and thus co-expressed with mammalian genes to identify the location of the protein encoded by the GFP-tagged gene in the mammalian cell. Such methods are useful tools for biologists in many areas of research, including those who study the mechanisms of human and other diseases or fundamental biological processes in eukaryotic or prokaryotic cells.

To date the broadest application of GMO technology is patent-protected food crops which are resistant to commercial herbicides or are able to produce pesticidal proteins from within the plant, or *stacked trait* seeds, which do both. The largest share of the GMO crops planted globally are owned by Monsanto Company, according to the company. In 2007, Monsanto's trait technologies were planted on 246 million acres (1,000,000 km^2) throughout the world, a growth of 13 percent from 2006.

In the corn market, Monsanto's triple-stack corn – which combines Roundup Ready 2 weed control technology with YieldGard Corn Borer and YieldGard Rootworm insect control – is the market leader in the United States. U.S. corn farmers planted more than 17 million acres (69,000 km^2) of triple-stack corn in 2007, and it is estimated the product could be planted on 45 million to 50 million acres (200,000 km^2) by 2010. In the cotton market, Bollgard II with Roundup Ready Flex was planted on nearly 3 million acres.

Rapid growth in the total area planted is measurable by Monsanto's growing share. On January 3, 2008, Monsanto Company (MON.N) said its quarterly profit nearly tripled, helped by strength in its corn seed and herbicide businesses, and raised its 2008 forecast.

According to the International Service for the Acquisition of Agri-Biotech Applications (ISAAA), of the approximately 8.5 million farmers who grew biotech crops in 2005, some 90% were resource-poor farmers in developing countries. These include some 6.4 million farmers in the cotton-growing areas of China, an estimated 1 million small farmers in India, subsistence farmers in the Makhathini flats in KwaZulu Natal province in South Africa, more than 50,000 in the Philippines and in seven other developing countries where biotech crops were planted in 2005.. ISAAA estimated that by 2008, 13.3 million farmers were growing GM crops, including 12.3 million in developing counties,. These comprised 7.1 million in China (Bt cotton), 5.0 million in India (Bt cotton), and 200,000 in the Philippines.

"The Global Diffusion of Plant Biotechnology: International Adoption and Research in 2004", a study by Dr. Ford Runge of the University of Minnesota, estimates the global commercial value of biotech crops grown in the 2003–2004 crop year at US$44 billion.

In the United States the United States Department of Agriculture (USDA) reports on the total area of GMO varieties planted. According to National Agricultural Statistics Service, the States published in these tables represent 81-86 percent of all corn planted area, 88-90 percent of all soybean planted area, and 81-93 percent of all upland cotton planted area (depending on the year). USDA does not collect data for global area. Estimates are produced by the International Service for the Acquisition of Agri-biotech Applications (ISAAA) and can be found in the report, Global Status of Commercialized Transgenic Crops: 2007. Transgenic animals are also becoming useful commercially. On 6 February 2009 the U.S. Food and Drug Administration approved the first human biological drug produced from such an animal, a goat. The drug, ATryn, is an anticoagulant which reduces the probability of blood clots during surgery or childbirth. It is extracted from the goat's milk.

Detection

Testing on GMOs in food and feed is routinely done by molecular techniques like DNA microarrays or qPCR. The test can be based on screening elements or event-specific markers for the official GMOs (like Mon810, Bt11, or GT73). The array-based method combines multiplex PCR and array technology to screen samples for different potential GMOs, combining different approaches (screening elements, plant-specific markers, and event-specific markers). The qPCR is used to detect specific GMO events by usage of specific primers for screening elements or event-specific markers.

To avoid any kind of false positive or false negative testing outcome, comprehensive controls for every step of the process is mandatory. A CaMV check is important to avoid false positive outcomes based on virus contamination of the sample.

TRANSGENIC MICROBES

Bacteria were the first organisms to be modified in the laboratory, due to their simple genetics. These organisms are now used for several purposes, and are particularly important in producing large amounts of pure human proteins for use in medicine. Genetically modified bacteria are used to produce the protein insulin to treat diabetes. Similar bacteria have been used to produce clotting factors to treat haemophilia, and human growth hormone to treat various forms of dwarfism. These recombinant proteins are safer than the products they replaced, since the older products were purified from cadavers and could transmit diseases. Indeed the human-derived proteins caused many cases of AIDS and hepatitis C in haemophilliacs and Creutzfeldt-Jakob disease from human growth hormone.

For instance, the bacteria which cause tooth decay are called *Streptococcus mutans*. These bacteria consume leftover sugars in the mouth, producing lactic acid that corrodes tooth enamel and ultimately causes cavities. Scientists have recently modified *Streptococcus mutans* to produce no lactic acid. These transgenic bacteria, if properly colonized in a person's mouth, could reduce the formation of cavities. Transgenic microbes have also been used in recent research to kill or hinder tumours, and to fight Crohn's disease. Genetically modified bacteria are also used in some soils to facilitate crop growth, and can also produce chemicals which are toxic to crop pests.

TRANSGENIC ANIMALS

Transgenic animals are used as experimental models to perform phenotypic tests with genes whose function is unknown. Genetic modification can also produce animals that are susceptible to certain compounds or stresses for testing in biomedical research. Other applications include the production of human hormones such as insulin. In biological research, transgenic fruit flies (*Drosophila melanogaster*) are model organisms used to study the effects of genetic changes on development.

Fruit flies are often preferred over other animals due to their short life cycle, low maintenance requirements, and relatively simple genome compared to many vertebrates. Transgenic mice are often used to study cellular and tissue-specific responses to disease.

This is possible since mice can be created with the same mutations that occur in human genetic disorders, the production of the human disease in these mice then allows treatments to be tested. In 2009 scientists in Japan announced that they had successfully transferred a gene into a primate species (marmosets) and produced a stable line of breeding transgenic primates for the first time. It is hoped that this will aid research into human diseases that cannot be studied in mice, for example Huntington's disease and strokes.

Cnidarians such as *Hydra* have become attractive model organisms to study the evolution of immunity. For analytical purposes an important technical breakthrough was the development of a transgenic procedure for generation of stably transgenic hydras by embryo microinjection.

Transgenesis in fish with promoters driving an overproduction of "all fish" growth hormone has resulted in dramatic growth enhancement in several species, including salmonids, carps and tilapias. These fish have been created for use in the aquaculture industry to increase the speed of development and potentially, reduce fishing pressure on wild stocks. None of these GM fish have yet appeared on the market, mainly due to the concern expressed among the public of the fish's potential negative effect on the ecosystem should they escape from fish farms.

Gene Therapy

Gene therapy, uses genetically modified viruses to deliver genes that can cure disease into human cells. Although gene therapy is still relatively new, it has had some successes. It has been used to treat genetic disorders such as severe combined immunodeficiency, and treatments are being developed for a range of other currently incurable diseases, such as cystic fibrosis, sickle cell anemia, and muscular dystrophy.. Current gene therapy technology only targets the nonreproductive cells meaning that any changes introduced by the treatment can not be transmitted to the next generation. Gene therapy targeting the reproductive cells-so called "Germ line Gene Therapy"-is very controversial and is unlikely to be developed in the near future.

Transgenic Plants

Transgenic plants have been engineered to possess several desirable traits, including resistance to pests, herbicides or harsh environmental conditions, improved product shelflife, and increased nutritional value. Since the first commercial cultivation of genetically modified plants in 1996, they have been modified to be tolerant to the herbicides glufosinate and glyphosate, to be resistant to virus damage as in Ringspot virus resistant GM papaya, grown in Hawaii, and to produce the Bt toxin, a potent insecticide.

Bt-maize is a corn that has been genetically modified by splicing the toxin-producing gene from bacteria into the DNA sequence of the corn in order to sicken or kill insects that try to consume it. While some genetically modified crops are more nutritious because they contain these extra vitamins and minerals, they will not cure all of the malnutrition-related ailments in the world and should only be a supplement to a balanced diet.

Cisgenic Plants

Genetically modified sweet potatoes have been enhanced with protein and other nutrients, while golden rice, developed by the International Rice Research Institute, has been discussed as a possible cure for Vitamin A deficiency. In reality, customers would have to eat twelve bowls of rice a day in order to meet the recommended levels of Vitamin A. In January 2008, scientists altered a carrot so that it would produce calcium and become a possible cure for osteoporosis; however, people would need to eat 1.5 kilograms of carrots per day to reach the required amount of calcium.

The coexistence of GM plants with conventional and organic crops has raised significant concern in many European countries. Since there is separate legislation for GM crops and a high demand from consumers for the freedom of choice between GM and non-GM foods, measures are required to separate foods and feed produced from GMO plants from conventional and organic foods. European research programmes such as Co-Extra, Transcontainer and SIGMEA are investigating appropriate tools and rules. At the field level, biological containment methods include isolation distances and pollen barriers.

CONTROVERSY

Biological Process

The use of GMOs has sparked significant controversy in many areas. Some groups or individuals see the generation and use of GMO as intolerable meddling with biological states or processes that have naturally evolved over long periods of time, while others are concerned about the limitations of modern science to fully comprehend all of the potential negative ramifications of genetic manipulation.

Foodchain

The safety of GMOs in the foodchain has been questioned, with concerns such as the possibilities that GMOs could introduce new allergens into foods, or contribute to the spread of antibiotic resistance.

Although scientists have assured consumers of the safety of these types of crops, consumption has been discouraged in many countries by food and environmental activist groups who protest GM crops, claiming they are unnatural and therefore unsafe. This has led to the adoption of laws and regulations that require safety testing of any new organism produced for human consumption.

Trade with Europe and Africa

In response to negative public opinion, Monsanto announced its decision to remove their seed cereal business from Europe, and environmentalists crashed a World Trade Organization conference in Cancun that promoted GM foods and was sponsored by Committee for a Constructive Tomorrow (CFACT).

Some African nations have refused emergency food aid from developed countries, fearing that the food is unsafe. During a conference in the Ethiopian capital of Addis Ababa, Kingsley Amoako, Executive Secretary of the United Nations Economic Commission for Africa (UNECA), encouraged African nations to accept genetically modified food and expressed dissatisfaction in the public's negative opinion of biotechnology.

AGRICULTURAL SURPLUSES

Patrick Mulvany, Chairman of the UK Food Group, accused some governments, especially the Bush administration, of using GM food aid as a way to dispose of unwanted agricultural surpluses. The UN blamed food companies and accused them of violating human rights, calling on governments to regulate these profit-driven firms. It is true that the acceptance of biotechnology and genetically modified foods will also benefit rich research companies and could possibly benefit them more than consumers in underdeveloped nations.

LABELLING

While some groups advocate the complete prohibition of GMOs, others call for mandatory labelling of genetically modified food or other products. Other controversies include the definition of patent and property pertaining to products of genetic engineering.

UNDERDEVELOPED NATIONS

Some groups believe that underdeveloped nations will not reap the benefits of biotechnology because they do not have easy access to these developments, cannot afford modern agricultural equipment, and certain aspects of the system revolving around intellectual property rights are unfair to undeveloped countries. For example, The CGIAR (Consultative Group of International Agricultural Research) is an aid and research organization that has been working to achieve sustainable food security and decrease poverty in undeveloped countries since its formation in 1971.

In an evaluation of CGIAR, the World Bank praised its efforts but suggested a shift to genetics research and productivity enhancement. This plan has several obstacles such as patents, commercial licenses, and the difficulty that third world countries have in accessing the international collection of genetic resources and other intellectual property rights that would educate them about modern technology. The International Treaty on Plant Genetic Resources for Food and Agriculture has attempted to remedy this problem, but results have been inconsistent. As a result, "orphan crops", such as tef, millets, cowpeas, and indigenous plants, are important in the countries where they are grown, but receive little investment.

PRIVATE INVESTMENTS

The development and implementation of policies designed to encourage private investments in research and marketing biotechnology that will meet the needs of poverty-stricken nations, increased research on other problems faced by poor nations, and joint efforts by the public and private sectors to ensure the efficient use of technology developed by industrialized nations have been suggested. In addition, industrialized nations have not tested GM technology on tropical plants, focusing on those that grow in temperate climates, even though undeveloped nations and the people that need the extra food live primarily in tropical climates. Many European scientists are disturbed by the fact that political factors and ideology prevent unbiased assessment of the GM technology in some EU countries, with a negative effect on the whole community.

TRANSGENIC ORGANISMS

Another important controversy is the possibility of unforeseen local and global effects as a result of transgenic organisms proliferating. The basic ethical issues involved in genetic research are discussed in the article on genetic engineering.

Some critics have raised the concern that conventionally-bred crop plants can be cross-pollinated (bred) from the pollen of modified plants. Pollen can be dispersed over large areas by wind, animals and insects. In 2007, the U.S. Department of Agriculture fined Scotts Miracle-Gro $500,000 when modified genetic material from creeping bentgrass, a new golf-course grass Scotts had been testing, was found within close relatives of the same genus (*Agrostis*) as well as in native grasses up to 21 km (13 miles) away from the test sites, released when freshly cut grass was blown by the wind.

GM proponents point out that outcrossing, as this process is known, is not new. The same thing happens with any new open-pollinated crop variety—newly introduced traits can potentially cross out into neighbouring crop plants of the same species and, in some cases, to closely related wild relatives. Defenders of GM technology point out that each GM crop is assessed on a case-by-case basis to determine if there is any risk associated with the outcrossing of the GM trait into wild plant populations.

The fact that a GM plant may outcross with a related wild relative is not, in itself, a risk unless such an occurrence has negative consequences. If, for example, a herbicide resistance trait was to cross into a wild relative of a crop plant it can be predicted that this would not have any consequences except in areas where herbicides are sprayed, such as a farm. In such a setting the farmer can manage this risk by rotating herbicides. The European Union funds research programmes such as Co-Extra, that investigate options and technologies on the coexistence of GM and conventional farming. This also includes research on biological containment strategies and other measures, to prevent outcrossing and enable the implementation of coexistence.

If patented genes are outcrossed, even accidentally, to other commercial fields and a person deliberately selects the outcrossed plants for subsequent planting then the patent holder has the right to control the use of those crops. This was supported in Canadian law in the case of Monsanto Canada Inc. v. Schmeiser.

"TERMINATOR" AND "TRAITOR"

An often cited controversy is a "Technology Protection" technology dubbed 'Terminator'. This yet-to-be-commercialized technology would allow the production of first generation crops that would not generate seeds in the second generation because the plants yield sterile seeds. The patent for this so-called "terminator" gene technology is owned by *Delta and Pine Land Company* and the United States Department of Agriculture. Delta and Pine Land was bought by Monsanto Company in August 2006. Similarly, the hypothetical Trait-specific Genetic Use Restriction Technology, also known as 'Traitor' or 'T-gut', requires application of a chemical to genetically modified crops to reactivate engineered traits. This technology is intended both to limit the spread of genetically engineered plants, and to require farmers to pay yearly to reactivate the genetically engineered traits of their crops. Traitor is under development by companies including Monsanto and AstraZeneca.

In addition to the commercial protection of proprietary technology in self-pollinating crops such as soybean (a generally contentious issue), another purpose of the terminator gene is to prevent the escape of genetically modified traits from cross-pollinating crops into wild-type species by sterilizing any resultant hybrids. The terminator gene technology created a backlash amongst those who felt the technology would prevent reuse of seed by farmers growing such terminator varieties in the developing world and was ostensibly a means to exercise patent claims.

Use of the terminator technology would also prevent "volunteers", or crops that grow from unharvested seed, a major concern that arose during the Starlink debacle. There are technologies evolving which contain the transgene by biological means and still can provide fertile seeds using fertility restorer functions. Such methods are being developed by several EU research programmes, among them Transcontainer and Co-Extra.

GENETICALLY MODIFIED CROPS

In October of 2000, the detection of the genetically engineered StarLink corn in Tacos and several other food products caught the attention of American consumers. During the last week of February, it was reported that most of the corn seeds ready for planting this year are contaminated with StarLink corn. Farmers and federal regulators, consequently, face a dilemma about what corn to plant. Americans are also increasingly aware of protests around the world objecting to genetically modified foods, identifying them as frankenfood or as environmentally unfriendly. These recent news items have re-ignited public debate over the technology of genetically altering crops. This and the subsequent two articles in future issues of this newsletter will attempt to address some common issues related to Genetically Modified Foods and the technology of genetic engineering. Part I will delve into the background of genetic engineering and the future of genetically altered foods; Part II will address issues related to the benefits and drawbacks of this technology; and Part III will delineate, in simple terms, the basics of how genetically modified organisms are created.

Selective plant breeding is not a new concept. Casual selection of observed desirable traits by our ancestors essentially tamed wild plants and made them suitable for agriculture. In the past, if pests devastated a field of crops and a few plants stayed alive and healthy, the seeds from these healthy plants were used to generate the next crop. Thus the beneficial factors that made the plants resistant were transferred to the next generation, making the new generation of crops slightly more resistant to the same pests. Such selections have been used for over 10,000 years, since the beginning of agriculture and have resulted in significant advances for humanity with increased yields, disease resistance and, overall, greater productivity. A good example is corn; the original crop was Teosinte with very small seeds and very few seeds in each cob. Over the centuries people selected for various traits thus improving the size of the corn kernel, the number of kernels in each cob, the stronger attachment of the kernels to the cob so that it could be harvested and so on, resulting in the corn that we are familiar with today.

It has since become evident that all traits, beneficial or otherwise, are conferred to living organisms by genes. Genes are small segments of the DNA (deoxyribonucleic acid) that code for specific proteins, which in turn regulate the various traits in all living organisms. There are thousands of genes in plants that regulate the activities governing the growth, stature, colour, and all other aspects of plant growth and survival. Selection through traditional breeding involved the transfer of numerous genes from one generation to the next, including the genes for the beneficial trait as well as the undesirable traits. It then took years of self-breeding and selecting to get a plant with both the normal characteristics as well as the beneficial trait desired.

The advent of genetic engineering greatly enhanced this process of transferring a beneficial trait into plants by directly transferring the gene/s responsible for the beneficial attribute. So in one generation, or one planting season, a plant can be created that is the same in all respects except the addition of the beneficial trait. Since genes in all living organisms code for similar proteins and properties, it is possible to transfer a gene from, say, one good corn variety to another corn variety or from fish to strawberry plants.

We can now define a genetically modified organism: It is any organism that has been modified by altering one or more genes by recombinant techniques. A recombinant technique is the method used to transfer a gene of interest from one organism to another. A genetically modified food therefore is any food that is produced from plants or animals that have been genetically altered using this method. The first genetically altered plant created was a tobacco plant with resistance to antibiotics in 1983. It was almost 10 years later when the first commercial genetically altered crop, a delayed ripening tomato, "Flavr-savr", was commercially released. This was not a commercial success, however, for reasons related to production and marketing strategies. This was soon followed by the release of several crops including Roundup Ready soy and corn. Corn and soy are the two most commonly used food crops that have been genetically altered. They have been primarily altered so that the plants can resist pests, diseases or chemicals used to destroy weeds in the field.

Such alterations that improve the health of the plant and potentially benefit farmers are commonly termed as input traits; that is farmers use fewer inputs to grow their crops, be it pesticide, herbicide or chemicals to prevent diseases. There are other alterations that are possible that alter the property of the oil or starch in the seeds. These are termed output traits; that is the seeds produced by the plant have altered properties either by way of improved yields, nutritional content or higher levels or quality of starch, proteins or oils.

Genetically altered foods are very prevalent, at least in the United States and the Western world. More than 60% of the foods we purchase from the supermarket today have ingredients derived from genetically modified crops. Most of these are either from corn or soybeans, which are the base for numerous ingredients manufactured for the food industry, including starch, oils, proteins and other ingredients.

Despite this prevalence, a recent USDA consumer focus group survey revealed that most consumers were unaware of the use of biotechnology in foods. Furthermore, the benefits of biotechnology were viewed as skewed towards producers and manufacturers, with little benefit to the consumer. There was also skepticism related to the long-term health effects and impact on environment. It is therefore essential that we disseminate the information about the technology of genetic engineering so that we can have an informed debate on the merits and shortcomings of this technology.

6

Plant Genome Structure

INTRODUCTION

The genomes of all eukaryotic species consist of single-copy, middle repetitive and high copy number sequences. To gain an understanding of the per centages of each of the classes within any particular species, a group of random clones can be hybridized to blots of plant DNA. One such study was performed by Zamir and Tanksley. They hybridized 50 random genomic clones to tomato DNA. Washing was performed at two stringencies. The following are the results with regards to copy number.

Clone Class	*Low Stringency Wash*	*High Stringency Wash*
Single copy clones	44%	78%
Multiple copy clones	46%	18%
Repetitive clones	10%	4%

The table shows that hybridization stringency has a significant effect upon the number of sequences to which a sequence hybridizes. At the higher stringencies, most of the clone recognized only a a single copy within the genome. What this also shows is that the tomato genome contains many sequences that are about 80% homologous, but fewer sequences that are highly homologous. Clearly similar sequences must have diverged by some mechanism during the evolution of the species. The authors also hybridized these 50 clones to filters containing tomato, related tomato species, eight Solanum species and one member of the Curcurbitaceae family.

As a control,single copy tomato cDNA clones were also hybridized to these clones. At moderate stringency, the cDNAs hybridized to all the tomato species, and most hybridized to the Solanum species. 80% of the clones hybridized to the related Curcurbitaceae species. In contrast, only 50% of the random clones hybridized to related tomato and Solanum species, and only 10 % hybridized to the Curcurbitaceae species. The principal conclusion that can be drawn from these hybridizations is that the random sequences (as represented by the random genomic clones) are evolving faster than the single-copy sequences. Why? At the higher stringency, homology to distant related species was reduced. Therefore these sequences have undergone greater divergence. This evidenced by the fact that 0/5 of the repetitive sequences hybridized to tobacco, whereas 20/45 of the single and multiple copy clones hybridized to tobacco.

EVOLUTIONARY RELATEDNESS

The typical experiment begins with a large collection of cultivars or species. These samples are then analysed by hybridization with RFLP clones or more recently by the PCR-RAPD amplification. The similarity of each pair of samples is measured by calculating the number of common bands or amplification products. One estimator was developed by Nei and L. The formula is:

$$F = 2nXY/(nX + nY)$$

where nX and nY are the total number of fragments for sample X and Y and nXY is the number of fragments shared by the samples. Data is normally collected for a relatively large number of hybridizations or PCR-RAPD amplifications. The similarity data is then used in a cluster analysis to develop dendrograms which show the molecular relatedness of the species.

These types of analyses can:

- Provide independent support to previous phylogenetic and evolution hypotheses.
- Identify gene pools within a genus.
- Estimate genetic diversity within a genus.
- Provide necessary data to select appropriate parents for a molecular mapping project.

PLANT GENOME ORGANIZATION AND STRUCTURE

Eukaryotic genomes are much more complex than prokaryotic genomes. And further, plant genomes are more complex than other eukaryotic genomes. Prior to the development of recombinant DNA technology genomes, were analysed by reassociation kinetics techniques. Reassociation kinetic experiments are performed by melting DNA and allowing it to reanneal upon itself or with another population of either DNA or RNA molecules. The kinetics of the reassociation provide data that can be used to analyse the overall structure, evolution and expression of genomes.

The most common method of denaturing duplex DNA is by heating to 100°C. But how can we monitor this denaturation and subsequent renaturation? The most common method is by measuring the absorbance change in the ultraviolet region at 260 nm. The important property is that melted, single-stranded DNA absorbs about 40% more at 260 nm than duplex DNA. If we slowly heat DNA the absorbance will increase dramatically over a short range of temperatures. The mid-point of this transition is called the melting temperature or Tm. Under physiological conditions, the Tm usually lies in the range of 85-95°C. Thus, without altering the cellular conditions, the duplex DNA is stable in the cell. The exact temperature that a particular DNA melts depends on several parameters. The GC content is important because GC base pairs have three hydrogen bonds compared to the two for AT base pairs. Thus, the higher the GC content the higher the Tm.

Table. Effect of GC Content on Tm

%GC	*Melting Temperature*
40	87°C
60	95°C

DNA denaturation is reversible, and the reversible process is called renaturation. This process requires that the temperature be lowered gradually.

As this lowering of temperature proceeds the following occurs:

- Single-stranded molecules randomly encounter each other.
- Short complementary stretches of duplex DNA form.
- The DNA then zips back together to form the original structure.

Obviously if a single molecule is denatured it should be able to reform completely. But if two separate molecules are denatured together, for example wheat and barley DNA, then complementary regions between the two DNAs will be able to form duplex DNA. The ability of two molecules to renature is called hybridization.

Hybridization: The pairing of complementary nucleic acids. Performing hybridization experiments in a solution is called liquid hybridization. We have already discussed the related procedure called filter hybridization.

COMPARATIVE GENOME MAPPING

Often the clones used to identify RFLPs in one species can be used in a second species. These heterologous probes can then be ued to develop a RFLP map in the second species. Once the two species have been mapped, the relative evolution of the two species can be compared. Comparative mapping can identify inversions, translocation and duplications that have occurred. Genetic factors can also be assessed by comparing map distances of genes with conserved gene order in the two species.

A comparison of the related grass species sorghum and maize was made using maize clones. The diploid chromosome number of the two species is twelve. Many of the sorghum chromosomes contained regions from two of the maize chromosomes. This result may represent the ancestral duplication of chromosomal material. Furthermore, during maize evolution duplicate genes have also occurred to a greater extent than that seen in sorghum. Only nine inversions of gene order have been observed between the species. Maize and sorghum were next compared with regard to genetic distance. Conserved gene orders were compared, and these linkages measured 862 cM in maize and 835 cM in sorghum. Therefore, it can be concluded that since the divergence of sorghum and maize, large chromosomal changes have not occurred and much of the recombination distance has been maintained.

The same type of analysis was performed with two more distant species, tomato and pepper. Little linkage conservation appears to have been maintained since the divergence of these two species, even though the chromosome number has been conserved. The largest conserved linkage block is a 63 cM block of tomato chromosome two that was located on chromosome I of pepper. Some chromosomes of pepper contained six distinct regions of the tomato genome (pepper chromosome X). These chromosomal breakages and rearrangements are not centromeric in nature, suggesting that evolution involved breakage throughout the genome. Because duplicated regions were found in pepper, it was concluded that the gene duplications events within pepper occurred after divergence from tomato. Linkage distances of conserved gene orders were quite similar.

PHYSICAL AND GENETIC DISTANCES

All distances found on a linkage maps are the product of genetic recombination and therefore are considered to be genetic distances. Two pairs of loci that are genetically the same distance apart may not be physically the same distance apart because of suppressed recombina-tion in the region between two of the loci. Those loci where suppression of recombination does occur will physically be farther apart. RFLP hybridizations to large fragments of DNA are necessary to accurately gauge the physical distance.These experiments are similar to other Southern hybridizations.

The DNA is cut with restriction enzymes first. The choice of enzymes is important because you want to generate large fragments. For these experiments you want to cut the DNA with enzymes that recognize 8 nucleotide sequences in the target DNA. Because these sites will be rare (every 65,536 bases on average) the digestion products will range from several hundred kilobase to several megabases. Because the normal gel systems do not separate large fragments very easily, it is necessary to run pulsed-field-gel-electrophoresis. This technique was first applied to the separation of yeast chromosomes and was quickly applied to large-scale mapping experiments.

Physical and genetic distances are then resolved by hybridizing clones which define closely linked genetic markers, to DNA that has been cut with rare cutting enzymes. What you are searching for is co-hybridization of the two clones to the same restriction fragment. If two clones hybridize to the same fragment, then the maximum distance between those two clones is the size of the restriction fragment. Because you already know the genetic distance between the loci, you can correlate the genetic and physical distances in this region. The following table give the relationship between the physical and genetic distance in three species.

Species	*Kilobases/centimorgan*
Arabidopsis	139
Tomato	510
Corn	2140

But remember, these values are estimates of a single region of the genome of each species, and it is quite common to see that two regions of the same species have different amount of DNA per genetic distance. Correlations between physical and genetic distances have also been performed using deletion stocks in wheat.

A series of deletions stocks of a specific wheat chromosome were analysed with a probes known to hybridize to that chromosome. "Each locus was assigned to the chormosome region between the breakpoint of the largest deletion where the band was present and the next larger deletion where the band was absent." (Werner *et al.* PNAS 89:11307–11311) The authors noted large discprencies between the physical and genetic maps of chromosomes 7B and 7D. Two loci that are located near the centromere of 7B are 7 cM apart genetically, yet the distance between the two loci spans 25% of the chromosome. Another region at the distal region of long arm of 7B which accounts for about 15 % of the chromosome is 91 cM long genetically. This points out the difference of recombination that can occur within a single chromosome.

LINKAGE DRAG

One goal of plant breeding is to introduce a gene from a donor parent to improve a cultivar for a specific trait. For example, a wild germplasm could be used as a source of disease resistance. At the same time the breeder does not want to carry any of the other genes from the wild germplasm that might reduce the a agronomic fitness of the cultivar.

The backcross method of plant breeding is one manner in which the introduction of a specific gene is accomplished. One genetic feature though of backcross breeding is linkage drag. This refers to the reduction in fitness in a cultivar due to deleterious genes introduced along with the beneficial gene during backcrossing. Molecular makers offers a tool in which the amount of wild or alien DNA can be monitored during each backcross generation.

First, though, lets look at the amount of alien DNA that can be maintained after a backcrossing programme. The accompanying figure handed out in class shows the chromosomal region around the Tm-22 allele in a number of tomato cultivars. This allele was introduced from the wild tomato species L. peruvianum, and it provides resistance to tobacco mosaic virus. Large variation in the amount of introgressed DNA was observed. For example, Craigella-Tm-22 contains 51 cM of wild DNA, whereas Vendor-Tm-22 and Nova-Tm- 22 each contain about 8 cM of introgressed DNA.

Backcrossed lines could be developed rapidly in which only a small amount of foreign DNA is linked to the gene of interest. After the first backcross all lines with the gene of interest could be screened with an RFLP that is 1 cM away. Those lines in which a crossover occurred at this marker would be selected. Then those lines would be crossed to the recurrent parent.

The progeny from these crosses would then be scored for a second marker 1 cM away on the other side of the gene. Again crossovers progeny could be selected. Thus, in only two generations the amount of wild DNA could be reduced to 2 cM. The key to this procedure is to have markers that are polymorphic between the two parents that are closely linked to the gene of interest. Young and Tanklsley applied these principles to the reduction of L. peruvianum DNA in Craigella-Tm-22. The authors were able to reduce the amount of DNA on one side of the Tm-22 allele from 47 to 7 cM in one generation by selecting for crossovers at the CD32A locus.

QUANTITATIVE TRAIT LOCUS

Traditional quantitative genetic research defined a quantitative trait in terms of variances. The total phenotypic was first partitioned into genetic and environmental variances. The genetic variance could then be further divided into additive, dominance and epistatic effects. From this information it was then possible to estimate the heritability of the trait and predict the response of the trait to selection. It was also possible to estimate the minimum number of genes which controlled the trait.

Mapping markers linked to QTLs identifies regions of the genome that may contain genes involved in the expression of the quantitative trait. But what functions could these genes be encoding. To answer this question we should consider a trait such as yield. What types of qualitative genes (genes inherited as simple genetic factors) could be involved in the expression of yield? The first event required for yield is meiosis.

Therefore any gene that is involved in gamete formation could potentially be considered a QTL. Any of the genes involved in the protein and carbohydrate biosynthetic pathways could also affect the final yield of a plant and could also be considered to be QTLs. As we saw above, the markers associated with a QTL each account for only a portion of the genetic variance. Likewise each of these genes of known function may only account for a portion of the final yield. An important question that can now be posed is whether any known genes map as QTLs.

Beavis *et al*. analysed four populations of maize and found molecular markers linked to plant height. No marker was consistently associated as a QTL with plant height in all four populations. Each of the ten maize chromosomes contained a marker linked to a QTL for at least one of the four populations. The authors further were able to demonstrate that a number of the QTLs identified by the molecular markers mapped to regions containing genes known to have a qualitative effect on plant height.

For example, on chromosome 9 the gene d3 resides within 10 cM of a plant height QTL. This gene is involved in gibberellic acid sensitivity. Mutants do not respond to the hormone and do not undergo the normal cell elongation. These mutants are phenotypically shorter than normal maize plants. The question that needs to be raised is if the QTL that was being identified by the molecular marker is actually the d3 gene. It could be possible that what is actually being measured by the marker is the linkage of the marker with the gene. The statistical analysis of quantitative traits provided valuable information for the plant breeder. Molecular analysis of quantitative traits now provides new tools, not only as selection tools for plant breeding, but as starting points for the cloning of these genes. These objectives could not have been realized without molecular markers.

ANALYSIS OF GENOMES BY REASSOCIATION EXPERIMENTS

If a DNA molecule is melted and allowed to reassociate, the complexity of the genome dictates the rate in which duplex DNA will form. If we consider a simple molecule that consists of alternating GCs, this molecule will be able to form a duplex quicker than a molecule that consists of repeating blocks of AGCT. As the number of different combinations of bases increases, the time required for complete duplex formation to occur will increase.

Renaturation, or duplex formation requires random collisions between two single-stranded molecules. This process follows second-order kinetics and is concentration dependent. We will not go through the derivation of the formula but the important parameter used to define a certain DNA is:$Cot^{1/2}$.

This value is defined as the amount of time required for one-half of the DNA to reanneal or form duplex DNA. The units for this parameter is moles of nucleotides per litre per second. The more complex the genome of interest, the longer it will take for like sequences to reanneal. Consequently, the $Cot^{1/2}$ will be larger. Thus in terms of reassociation kinetics complexity has a specific definition.

Complexity: The total length of different sequences. For example, E. coli is considered to have a complexity of 4.2 × 106 base pairs. What is the experimental procedure used to derive these values?

In general the procedure is:

- Shear the DNA to be analysed to a length of about 300 bp.
- Melt the DNA (usually in 0.12 M phosphate buffer) by boiling for 5 min.
- Quickly place at 60°C.

- Take aliquots at different time points. Separate single-stranded DNA from double-stranded DNA by hydroxyapatite. Measure the amount of DNA that is double-stranded by absorbance at 260 nm.
- Plot the amount that is single-stranded versus the Cot value. The Cot value is expressed in log equivalent. This plot depicts the Cot curve.

When this type of experiment is performed with eukaryotic DNA three components are usually seen. These components each reanneal with their own unique Cot½ value. The three components are termed the fast, intermediate, and slow components. Why do we see these three components? Eukaryotic genomes are characterized by sequences that are represented by different copy numbers. If a sequence is found many times in the genome, it will reanneal much quicker than those sequences that are found only once in the same genome. Thus the equivalent Cot curve for a eukaryotic genome will be different than a genome, such as E. coli, which only contains single copy sequences.

A comparison of the Cot value of each of these components with an E. coli standard allows us to derive the complexity of each component. The complexity of the slow component of the genome is greater than that for the other two components and is considered to represent the single copy portion of the genome. The complexity of the slow component can be used as a good estimate of the genome size.

The genome size will be the sum of the lengths of all the unique sequences. Using the example from Genes V - Lewin, p.664, the complexity of the slow component is 3×10^8 bp and the complexity of the intermediate component is 6×10^5. If we divide the complexity of the slow component into the intermediate component we get 2×10^{-3}. This demonstrates that the intermediate component contributes very little to the complexity of the genome. Therefore, the complexity of the slow or single copy portion of the genome can be considered equal to the genome size.

To derive the complexity of each component, it is necessary to run a standard, such as E. coli DNA, with each experiment. E. coli is considered to consist of only single- copy sequences. Let's say that in the experiment from which the Cot for each component was derived, the Cot½ value for E. coli was 4. Experimentally it was determined that the slow component comprised 45% of the total DNA. Therefore, if only that component was annealed, the Cot½ value would be 283 (630×0.45). That value is 71 (283/4) times slower than for E. coli.

Therefore the complexity of the slow component is 71 times that of E. coli or 3.0×10^8 ($71 \times 4.2 \times 10^6$). The complexity of the other components is derived similarly. Genome size is quite variable throughout the biological world and the genome size in plants shows the greatest variation of any kingdom in the biological world.

Table. Variation in Genome Size among Plants

Species	*kb/haploid*	*pg/haploid*
Arabidopsis	7×10^4	0.15
Lily	1×10^8	100.00

Conversion factor: 1 pg = 0.965 × 109 bp = 6.1×10^{11} daltons

One manner in which a genome can be described is by determining the distribution of fast, intermediate and slow components in the genome. For comparison purposes, what does the human genome look like? Distribution Sizes Among Components of the Human Genome

Component	*% of Genome*
Fast	6
Intermediate	38
Slow	50

The table Sequence Distribution of Selected Plant Species lists the different components from different plant species. As you can see, plant species exhibit a wide range of values for each of the components. The genome of Arabidopsis is essentially entirely single copy sequences (the repetitive sequences have been determined to be essentially all chloroplast DNA).

At the other extreme, pea and wheat genomes have only 10–20% single copy sequences. Reassociation kinetic experiments of polyploid species, such as bread wheat (Triticum aestivium) were unable to derive a component that displayed true single copy kinetics. Instead the slowest component appeared to act as if it consisted of copies represented three times.

This result is consistent with the current hypothesis that bread wheat was developed from the introgression of three diploid wheat species. Genomic analysis suggest that this hypothesis is correct since the slowest component appears to consist of sequences represented three times.

REPEATED SEQUENCES

The intermediate and fast components are composed of sequences that are found many times in the genome. These sequences are called repetitive sequences and can vary in size from a 100 bp to 1000 bp or more. Furthermore, these sequences have undergone sequence divergence by the addition or deletion of sequences or by changes in the base pair sequence. Thus, the repeated sequences themselves show some divergence. An example of a highly repetitive sequence is the repeat found to be associated with the knob heterochromatin of corn. It ranges from 3-5 × 10^5 copies on a small knob to 1 × 10^6 on the large knobs. This sequence is unique to knobs and is not found associated with any other heterochromatic regions of corn. An example of a functional repeated sequence in plants is the corn storage proteins, zeins. Two major classes of zeins exist, the 22 and 19 kd classes. Sequence analysis has shown that both classes have the same structure. The only difference between the 22 and 19 kd class is the repeat unit. The 22 kd class has 8 repeat units and the 19 kd class has 7 repeat units. Estimates have been made of the number of copies of these genes and 30-50 copies of the 22 kd class are found in the corn genome. Thus, the repeat unit would be represented 240-400 times in the genome.

ORGANIZATION OF SINGLE-COPY SEQUENCES

Single-copy sequences are interspersed throughout the plant genome. These sequences are bounded by repeat sequences. The length of the single-copy regions varies widely among plant species.

In general, two types of arrangements are recognized:

- *Short Period Interspersion*: Single copy sequences of 300-1200 bp are interspersed as islands among short lengths of repeat sequences
- *Long Period Interspersion*: Single copy sequences of 2000-6000 bp are interspersed as islands among repeat sequences

How can the interspersion type be determined? Reassociation kinetic experiments are performed with 300 nt long fragments. These experiments give a characteristic Cot curve that defines each of the components. But what happens if the fragment that is followed is of a longer length, for example 900 nt? This fragment could be considered to consist of three 300 nt fragments, and each fragment may be from any of the three components.

Indeed this is how these experiments may be performed. A tracer of a longer length (such as 900 nt) is prepared and radiolabelled, for example with H^3. This tracer is added to a normal Cot reaction in such a low concentration that the rate is not affected. The reaction is then allowed to proceed to a Cot value determined from a normal experiment with only 300 nt fragments where the repetitive sequences have reannealed, but the single-copy sequences have not.

Then the amount of tracer which remains single-stranded is determined. What will be seen is a reduction in the amount of expected single-copy sequences. Why? If the 900 nt long fragment contains both single copy and repetitive sequences it will reanneal under these conditions as a repetitive sequence, and thus the amount of DNA that appears in the single-copy fraction will be reduced proportionally. Therefore, that single-copy sequence is interspersed with repetitive sequences. If for example the single copy fraction is determined to account for 40% of the genome when 300 nt fragments are used, and we calculate that the genome has 30% single-copy sequences when a 900 nt tracer is used, 10% (40%-30%) of the 900 nt fragments contain single-copy and repeat fragments and that 25% (10%/40%) of the single-copy DNA is interspersed with repetitive sequences at an interval of 300-900 nt.

A final concern is the size of the repeat units. This can be obtained by using a long tracer (5000 nt, for example) and reannealing to an appropriate

Cot such that only repeat sequences bind to the tracer. The product is then treated with an enzyme that cuts only single-stranded DNA. This enzyme is S1 nuclease. After digestion you obtain products that only contain repeat sequences. These products are then sized to give the average, mode and range of repeat sequences.

EVOLUTION OF REPEATED SEQUENCES IN CEREALS

The analysis of repeated sequences has also provided experimental evidence which supports the current model of cereal speciation. The following tables and the associated diagram detail the results. These show that once a repeat enters the lineage it remains.

Therefore group I repeats are found in all of the species because it is the most ancient repeat. Secondly, new repeats are added to the lineage and appear in all species which subsequently diverged from the lineage. Finally, species-specific repeats are formed in each species after that species has diverged from the common lineage. The distribution of the lineage specific repeat supports the speciation model.

ESTIMATING THE NUMBER OF EXPRESSED GENES

Reassociation kinetics can also estimate the number and abundance of expressed genes. These experiments are performed using high concentrations of RNA and tracer levels of either DNA or cDNA.

These experiments are analagous to DNA reassociation experiments except the units are expressed as moles of ribonucleotides per litre per second and the value is called a Rot.

If tracer amounts of DNA are allowed to reanneal only a portion of that DNA will form a duplex because not all the DNA will be expressed in the RNA population to which it is being hybridized.

Let's use the example in the Genes V to derive an estimate for the number of genes that are being expressed. In this experiment 1.35% of the DNA hybridized to the RNA.

Since RNA is single stranded the other anti-sense strand of the DNA would not have a partner with which it could hybridize. Thus actually 2.7% of the DNA is represented in this RNA population. If the genome size is 8.1 X 108 bp and the single-copy sequences represent 75% of the genome then we can estimate the complexity of the expressed DNA.

$0.027 \times 0.75 \times (8.1 \times 108$ bp$) = 1.7 \times 107$ bp If each gene is about 2000 base pair then the number of genes that is expressed is:

$$7 \times 107 \text{ bp} / 2000 = 8500 \text{ genes}$$

CHLOROPLAST GENOME ORGANIZATION

All angiosperms and land plants have cpDNAs which range in size from 120-160 kb; three expceptions are:

Species	Size (kb)
N. accuminati	171
Duckweed	180
Geranium	217

All cpDNA molecules are circular and spinach is used as the basis for all comparisons. Very few repeat elements are found other than short sequences of less than 100 bp. The notable exception is a large (10-76 kb) inverted repeat section, which when present, always contains the rRNA genes. (Legumes such as pea do not contain this repeat.) For the majority of species, this repeat region is 22-26 kb in size. Finally,the genetic order of the ribosomal unit is conserved in all species:

$$16S - tRNA^{ile} - tRNA^{ala} - 23S - 5S$$

Two other features of chloroplast DNA is described. First it was shown to that it can exist in in two orientations This implies that the molecule can undergo an isomerization event. Second is has been shown that spinach, corn, tomato and pea can all exist as multimers (PNAS 86:4156, June 1989).

Multimer	Relative Abundance	Per cent
Monomer	1	67.5
Dimer	1/3	22.5
Trimer	1/9	7.5
Tetramer	1/27	2.5

Because photosysnthesis is the primary function of the chloroplast it is not surprising that the chlroplast genome contains genes which encode for proteins that are involved in that process.

Reaction	Function
Dark Reactions	rbcS (nuclear encoded)
	rbcL (chloroplast encoded)
Light Reactions	apoproteins for PSI andPSII
	cytochrome b6
	cytochrome f
	6 of 9 ATPase subunits
	cab, LHC proteins (nuclear encoded)
	plastocyanin (nuclear encoded)
	ferredoxin (nuclear encoded)
Other	19/60 ribosome binding proteins
	translation factors
	RNA polymerase subunits
	tRNA and rRNA genes

Atrazine resistance is apparantley mediated through the psbA gene sequences of the 32 kd protein which is encoded by cpDNA. DNA sequence analysis revealed the following amino acid changes that are thought to be important.

Species	*AA#*	*Susceptible*	*Resistant*
Blue green algae	264	Ser (TCG)	Ala (GCG)
Chlamydomonas	264	Ser (TCT)	Ala (GCT)
Solanum nigrum	264	Ser (AGT)	Gly (GGT)
Amaranthus	228	Ser (AGT)	Gly (GGT)

Evolutionary Changes of cpDNA

- The majority of changes are small insertions and deletions of 1-106bp; significantly, a few length mutations of 50-1200 bp are clusted in "hot spots".
- The largest deletion occured in pea where an entire rRNA cluster is lost.
- The most common evolutionary change is in gene order. Small changes in the gene order occur, especially in the algae, but inversions have generated large scale order changes:
  - *Legumes*: About 50 kb inversion brought rbcL closer to psbA
  - *Wheat*: About 25 kb inversion brought atpA closer to rbcL

MITOCHONDRIAL GENOME ORGANIZATION

In comparison to the chloroplast genome, the size of the mitochondrial genome is quite variable.

Species	*Size (kb)*
Oenothera	195
Turnip	218
Corn	570
Muskmelon	2400

Further, in comparison to the mitochondrial genomes of other species the size is quite large and variable. For example, animal mitochondrial genomes range in size form 15-18 kb, and fungi mitochondrial genomes range form 18-78 kb. Plants may code for more proteins than with species. For example, genes for ribosomes, subunits I and II of cytochrome oxidase and ATPase subunits are located on the mitochondrial genomes of plants.

When DNA from corn mitochondria was investigated with EM, several circular molecules of different sizes were detected. Once the genome was mapped it became apparent that a mechanism existed to generated these circles of different sizes. It is now understood how these molecules arise. First, lets look at the simple situation of turnip. Two direct repeats undergo intramolecular recombination to give the two smaller molecules:

$$218 \text{ kb} \rightarrow 135 \text{ kb} + 83 \text{ kb}$$

The mitochondrial genome of corn undergoes the same type of recombination, but the events are more complex. First, the master circles can be subdivided into two major subgroups:

$$570 \text{ kb} \rightarrow 488 \text{ kb} + 82 \text{ kb}$$
$$570 \text{ kb} \rightarrow 503 \text{ kb} + 67 \text{ kb}$$

The second group of molecules are still labile and can produce several other subpopulations. Further two subgenomic circles can unite to form a larger circle. This variability is possible because corn has 10 repeats with which intramolecular recombination can occur.

Species	*Master Circle Size (kb)*	*Sub-genomic Circle Size (kb)*	*Repeat Size (kb)*
Turnip	218	135 + 83	2
Cauliflower	217	172 + 45	?
Black Mustard	231	135 + 96	7
White Mustard	208	none	none
Radish	242	139 + 103	10
Spinach	327	234 + 93	6

Introns have been located in the cytochrome oxidase subunit II gene (the cytochrome complex consists of 3 mitochondrial and 4 nuclear encoded genes). This gene contains one intron in rye, corn, wheat, rice, and carrot, but for other species such as Oenothera, broad bean, cucumber the gene has no intron.

Promiscuous DNA

Stern and Lonsdale (1982) hybridized mtRNA to a SstII digest of maize mt DNA and found that it hybridized to fragments known not to contain mt rRNA genes. The question of interest was - what was it hybridizing to? They next looked at a cosmid clone of corn mtDNA that hybridized to the mt RNA and found that it hybridized to a RNA molecule of the size of the cp 16S RNA gene. How could this have happened?

They next mapped the clone and compared it to the map of the corn cpDNA and found that the clone map was almost congruent with that of the the cpDNA 16S RNA region. This mapping showed that the two maps were nearly identical over a 12 kb region of DNA. These results suggest that cpDNA had been transferred to the mitochondrial genome.

The observation that organelle DNA was found in other DNA compartments of the cell was extended by other researcher. Stern and Palmer looked at corn, mung bean, spinach and pea and found extensive evidence of cpDNA/mtDNA homology.

These observations were extended to other DNA locations in the plant cell. Kemble *et al* (1983) demonstrated that mitochondrial DNA sequences are located in the nucleus of corn. Scott and Timmis (1984) showed that cpDNA sequences are found in the nuclear DNA.

RNA EDITING

The Central Dogma of Molecular Genetics states that the information that is found in DNA is used to produce mRNA molecules that are instrumental in the production of proteins. Therefore, the information flows directly from DNA to protein, via the RNA intermediate molecule. Recently it has been discovered that the information that is contained in the DNA is not always found in the RNA products used to make proteins.

It has now been demonstrated that mitochondria and chloroplast contain the biochemcial machinery to alter the sequence of the final transcription product. This process is called RNA editing. This process was identified in the following manner. Sequence analysis of a number of cytochrome c oxidase subunit II genes from non-plant species revealed that a tryptophan residue was invariant at several locations in the final protein product. But sequence analysis of this gene in several plant species revealed arginine at those positions.

Since a single base pair change in the codons for the two amino acids could generate this change (CGG for UGG), it was suggested that CGG encoded for tryptophan and not arginine in plant mitochondria. (This is the only change in codon usage that has been suggested for plants and has been postulated for several other genes as well.) But this change in codon usage was not universal, that is some CGG codons actually specified arginine in the final protein product. Furthermore, no amino acyl tRNA that recognized CGG was found to be charged with tryptophan, a prerequisite if this codon specification was actually real.

The solution to this dilemma was found by sequencing the mRNA products for cytochrome oxidase subunit II genes. It was found that in the mRNA the cytosine residue had been changed (edited) to uridine at the sequence location where the invariant tryptophan residue is found. This changed the codon at that location to UGG which is recognized by a tRNA that carries the amino acid tryptophan. An analysis of three other plant mitochondrial genes where the same altered codon usage was predicted suggested that mRNA editing was also occurring at the codon and that a cytosine residue was edited to uridine.

This editing process has also been detected in protozoa and it remains to be determined if RNA editing is a widespread function in mitochondria. A final point that this editing function highlights is that the sequence that is found in the DNA is not entirely and faithfully represented in the final protein product.

Specific Features of RNA Editing

- Editing can occur in both mitochondria and chlorplasts
- To date, >300 different editing events have been detected in plant mitochondria.

- The vast majority of the events involve a C to U tansition. A few cases of U to C transitions have been reported. This suggests that the editing machnery can also carry out the reverse modification.
- RNA editing can modify from 0.8% to 5.8% of the nucleotides of a specific transcript.
- The RNA editing events appear to occur at random in the transcript.
- Both 5' and 3' non-coding regions of mRNAs have also been shown to be edited.
- Structural RNAs such as tRNAs and rRNAs do appear to be affeceted.
- Editing can convert a tryptophan codon to a arginine codon (CGG to UGG).
- Start AUG codons can be created from ACG threonine codons
- Stop codons can be created by editing CAG, CAA and CGA codons.
- The most frequent amino acid substitions derived from RNA editing are Pro to Leu, Ser to Leu and Ser to Phe.
- Plant mitochondria do not use the universal genetic code.

The primary benefit of RNA editing could be evolutionary conservation of protein structure. For example, bound copper is required for the funciton of cytochrome c oxidase subunit II (coxII). After editing, all amino acids #228 are converted to cysteine, an amino acid required for copper to bind. In all species except for plants, the coxII gene encodes for methionine at codon #235. In plants, this methionine is generated by RNA editing. These events suggest that this protein is under very strong structural and functional constraints.

FUNCTIONAL GENOMICS IN PLANT CELL BIOLOGY

Genomic resources have significantly impacted plant biology research in recent years. Cell biology has been further enabled by an ongoing revolution in visualization technologies. Using fluorescent proteins (FPs), we now have unprecedented views of cellular architecture, and we can study real-time dynamics of cell structure, function, and protein localization. To date, these technologies have been most widely used in Arabidopsis (Arabidopsis thaliana); however, the grasses provide a unique opportunity to study the underlying mechanisms and inter-related controls of cell growth, morphogenesis, and physiology in leading crop models.

Plants are photoautotrophs: The most obvious distinguishing characteristic of plants is the presence of chloroplasts, which enable them to photosynthesize. They are thus photoautotrophs. The presence of plastids also has implications for genetics, since plastid genomes can carry important traits, most of which are directly related to photosynthesis.

Cell walls constrain development: Plants have cell walls made of cellulose, fungi have chitinous cell walls, while animals do not have cell walls. Plants maintain their structural integrity through a combination of mechanisms: turgor, cell walls, and insoluble components such as lignin. Animals derive their structural integrity through endo- or exo-skeletons, muscle and connective tissue. Higher plants and animals have circulatory systems that allow for the distribution of nutrients and the removal of wastes. Fungi are constrained in size due to the lack of any comparable features.

- Another consequence of the cell wall has to do with the limitations it imposes on diffusion. With pores of 3.5 to 5.2nm, movement of molecules >15Kd is significantly impeded. Thus, plants must subsist on molecules of low molecular weight, and any intercellular signaling molecules that have to pass through the cell wall must also be small.
- *Plasmodesmata facilitate symplasty*: Plasmodesmata convert a plant from being just a collection of individual cells to a large, interconnected symplast. Although the passage of molecules through plasmodesmata is tightly controlled, this provides a means for intercellular communication that is not available in animal tissues.
- *Most cell volume is taken up by vacuoles*: Often 50%, to as much as 95% of the volume of a plant cell is taken up by vacuoles. The vacoule makes it possible for plant cells to become very large, without a correspondingly large increase in cytoplasmic volume. This is important because the cytoplasm is requires a fairly large investment, in terms of macromolecular synthesis and turnover. Additionally, by limiting the volume of the cytoplasm, the rates of intracellular diffusion are not slowed down. This in turn affects diffusion-dependent processes, such as signal transduction from the plasmalemma or cytoplasm to the nucleus, protein synthesis, protein transport, membrane synthesis, cytoplasmic streaming, and absorption and equilibration of nutrients, gases, and other small molecules.

LIFE HISTORY OF PLANTS

- *The meristem (germline)*: The germ line - Animals have specialized germline tissues, spermatogonia and oogonia, which diverge from somatic cells early in embryogenesis, and are largely insulated from selective pressures. In contrast, flowering plants have large portions of their undifferentiated tissue as meristem, tissue capable of growing vegetatively and differentiating (often in response to very specific environmental cues, such as day length) into reproductive organs which then produce germ cells.
- *Gametophyte and sporophyte generations*: In animals, the final meiotic products (spermatids and ova) are mostly metaboli-cally dormant, and actually shut down gene expression when they have completed

maturation. In fact, egg cells in animals typically have enough mRNA and rRNA pre-synthesized to last the zygote for several cell divisions after fertilization.

Flowering plants have a 2 part life cycle, a diploid (or greater) sporophyte stage, and a haploid gametophyte stage.

Female: Megaspore mother cell (2N) → megaspore (1N) → megagametophyte → egg

Male: Microspore mother cell (2N) → microspore (1N) → pollen grain → sperm

Thus plant germline (gametophyte) cells undergo a number of cell divisions in the haploid state, and, presumably, have to perform all of the normal metabolic housekeeping functions. An important implication of this is that recessive lethal alleles in animals will be carried into the zygote, and selection against these traits may be prevented by heterozygosity. In plants, a good number of recessive lethals (eg. metabolic pathways, cell cycle controls) would be weeded out during the haploid gametophyte generation. One piece of evidence that could corroborate this is that it is possible to produce haploid plants from some species (eg. canola, tobacco). Admittedly, these plants are usually pretty sickly, but the fact that you can do this at all may be a result of this weeding out of deleterious recessive alleles.

- *Totipotency*: Plant somatic cells remain totipotent throughout the life of the plant. Animal cells tend to become terminally differentiated early in development, and can undergo genetic rearrangements which destroy totipotency. Examples: selective loss of chromosomes from non-germline cells in some invertebrates; selective polytenization of parts of chromosomes; somatic amplification of rRNA genes; rearrangements within immunoglobulin genes in lymphocytes.

*To elaborate on this last exampl*e: In vertebrates, the immunoglobulin genes in lymphocytes undergo rearrangements of DNA in the regions encoding the hypervariable regions of antibody proteins. Since a different rearrangement occurs in each cell, the population of lymphocytes can differentiate into T-cells and B-cells, each of which will carry antibodies with unique specificties. These DNA rearrangements are irreversible. The totipotency of plant cells has important implications. First, it deprives plants of one, admittedly drastic, means of genetic regulation. More importantly, it has made the transformation of plants for experimental and genetic engineering purposes a great deal easier.

- *Differentiation without migration*: Animal cells must migrate by amoeboid motion through the embryo during development. Plant cells are constrained by their cell walls to stay where they are. Therefore, differentiation must be carried out without migration. This implies that there are very fundamental differences between plant and animal ontogeny. Is this inability of cells to migrate an evolutionary dead end for plants? Is this the key to why animals seem to have a greater potential for physiological complexity than plants?

- *Continuous programme of development*: The totipotency of plant cells, along with the relative lack of terminal differentiation during vegetative growth results in a continuous programme of differentiation and development throughout the lifetime of the plant. In animals, once an organ or tissue has formed, that's it. You could say that, while animals go through a single iteration of the developmental programme, plants can go through multiple iterations, usually in response to environmental stimuli. This can be viewed in its extreme as the abilty of new plants to be produced by vegetative propogation (eg. potatoes) without ever going through a gametophyte stage.
- *Shift between vegetative and reproductive growth*: In animals, germline cells become differentiated very early in development, and specialized sex organs are established early. In flowering plants, sporophytes grow vetatively for an extended period. The shift from vegetative to reproductive growth is triggered by environmental cues. It is during flowering that reproductive organs develop.
- *Kinetics of cell growth*: The same kinetic principles that govern bacterial cultures govern plant growth. In a sense, cells in an organ can be thought of as a culture.

Per centage growth per day as a function of age: The growth of a local population of cells is limited in part by the ratio of the mass of cells to vascular tissue.

Hypothesis: All the genetic programme needs to do is control the key points at which vascular tissue branches out, and the adjoining mesophyll and epidermal cells come along for the ride, filling out the shape, as it were.

Thus, things like the thickness of leaves, or the distance between veins are determined by diffusion limitations. Superficially, we're not just talking about growth as cell division, but there is also a component of elongation. Both these processes must be under the control of developmental regulators.

- *Developmental plasticity*: The plant body plan is more of an iterative loop, rather than a specific blueprint, as in animals. The specific location, and number or organs, is often environmentally determined:
  - Switch from vegetative to reproductive growth (flowering) does not necessarily occur at a specific stage of growth
  - Number of leaves or nodes
  - Tissues such as roots can grow in direction of nutrients; leaves and stems can grow towards light.
- *Localized senescence*:
  - leaves senesce and abscise, while branches persist. (One of the few analogues in animals might be a snake shedding its skin.)
  - biomass from senescent tissue can be resorbed and transported to other tissues eg. seeds.

- *Cell division not required for new cell identities*:
  - Mesophyll cells can differentiate into tracheary elements through a breaking down of the cytoplasm and vacuole, and thickening of the cell walls.

GLOBAL ANALYSIS OF GENE EXPRESSION

The experiments of Goldberg and others on global analysis of gene expression provided what is, even today, the most comprehensive look at the role of gene expression in development. The key problem with R0t analysis, as done in these studies, is that nothing is learned about which genes or pathways are important in each organ or tissue studied. We can estimate the complexities of the RNA populations, and even learn, for example, that gene expression in leaves is very similar to that in petals, but quite dissimilar to that in roots. However, not knowing the identities of the genes involved limits the usefulness of these findings.

Most of the history of molecular biology has been characterized by the use of cloned probes, typically in Northern blots, to measure gene expression one gene at a time. This is precise and specific, but still has important limitations. First, the number of genes that can be examined is limited for practical reasons. You have to have a clone for every gene you wish to study. Secondly, it is necessary to decide which genes to look at. Unfortunately, choosing which genes to study biases the experiment. Important genes might be missed, and the role of relatively minor genes could be overestimated. One of the unifying themes in the field of genomics is that by working on a large scale, you get a more comprehensive picture of the subject. We will look at several large scale approaches towards finding important genes in biological processes.

Gene Arrays

EST studies begin to tell us about the expression of the genome, but all they really tell us is that a gene is expressed in the cells from which the ESTs were derived. Gene array experiments give a more precise picture of the 'transcriptome', the set of transcribed genes, in specific developmental stages, or in response to treatments.

- *Gene array technology*: Measures mRNA levels for thousands of genes.
  - Any cell or tissue type.
  - At any point in development.
  - In response to any stimulus.

Gene arrays consist of hundreds or thousands of cDNAs spotted onto microscope slides (microarray) or nylon filters (macroarray). cDNAs are chosen from EST collections, so the sequences, and usually the identities of genes in the array are known. In a gene array experiment, an mRNA population is isolated from cells. The population is labeled by synthesizing complementary cDNAs using reverse transcriptase and labeled nucleotides. The resulting cDNA population is then hybridized to the array.

- *Gene* x: Strongly expressed; high abundance transcript
- *Fene* y: moderately expressed; medium abundance transcript
- *Gene* z: weakly expressed; low abundance transcript

Each transcript base pairs with the complementary DNA for its corresponding gene on the array. Signal strength is proportional to the abundance of each mRNA *Example*: cDNA probes were made from Human fibroblasts treated with serum (Cy5-dUTP) or serum-deprived cells (Cy3-dUTP). 3 replicate arrays were scanned, and the same regions from the replicate arrays are shown. Serum inducible genes appear green, while serum suppressible genes appear red. 1 - protein disulfide isomerase-related protein P5; 2 - IL-8 precursor; 3 - EST AA057170; 4- vascular endoghelial growth factor.

- *Two general classes of data*: Gene array studies tend to generate two different types of data. Studies in which two or more conditions are compared at a time generate discrete state data. Often it is critical to follow the expression of a gene over time after a treatment. In timecourse experiments, the expression of each gene in response to two or more treatments is measured over time. For example, in the timecourse at right, the solid blue and red dashed curves might represent the expression levels for a gene in response to two different drugs.

There is a whole family of problems in normalization of data and controlling for components of experimental variation. To put things into perspective, if the experiment was repeated 3 times, the timecourse above represents 2 treatments × 6 times × 3 replicates = 36 probes hybridized to 36 duplicate arrays to generate the data.

Although the data for each replicate are averaged, there is often a great deal of variation in the results, which can potentially negate any meaning. Therefore, extraordinary measures must be taken to minimize experimental variation at each step in the procedure, to minimize the overall variation. What we're ultimately trying to get from gene array experiments is expression patterns for each of the hundreds or thousands of genes in the array. By identifying genes whose expression patterns are similar, we can discover which groups of genes work in concert, in response to a given stimulus.

SEED DEVELOPMENT

During seed development, the embryo differentiates into two organ systems: the axis and the cotyledon. The axis contains root and shoot meristems that will give rise to the mature plant after seed germination. By contrast, the cotyledon is a terminally differentiated organ system that senesces after germination and is responsible for synthesizing andstoringfood reserves used by the germinating seedling.

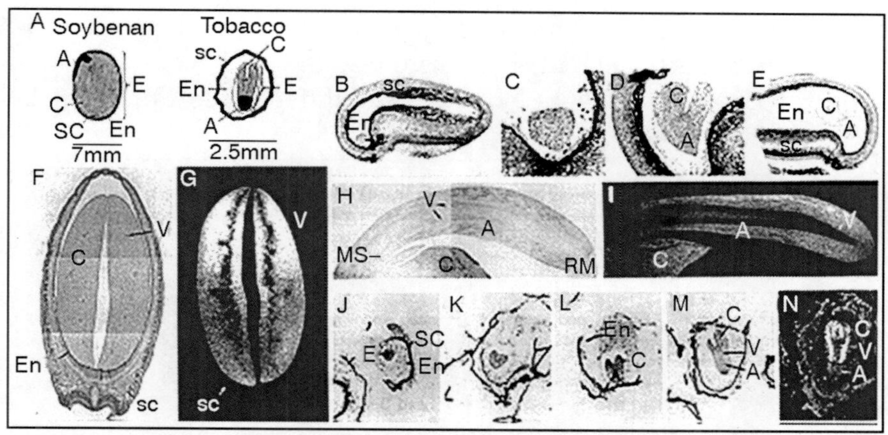

β -Conglycinin mRNA In Soybenan And Transgenic Tobacco.

β-conglycinin - storage protein, accumulates to high levels in soybean seeds. ~15 genes in soybean, clustered in two different chromosomal loci. 2-G - Dark-field microscopy shows silver grains as bright signal against dark background. β-conglycinin mRNA accumulates solely in the cotyledons (c) and not in surrounding embryonic tisssues (e). In the cotyledons, it accumulates in the storage parenchyma tissues and not in the vascular cylinder (v). 2J-M - development of tobacco seed. 2N - In-situ hybridization of transgenic tobacco expressing soybean beta-conglycinin.

- Similar expression patterns to soybean ie. only in cotyledon, non-vascular tissues
- Indicates that signals for tissue-specific expression in developing embryo must be well conserved.

Embryogenesis

Plant embryogenesis is the process that produces a plant embryo from a fertilised ovule by asymmetric cell division and the differentiation of undifferentiated cells into tissues and organs. It occurs during seed development, when the single-celled zygote undergoes a programmed pattern of cell division resulting in a mature embryo. A similar process continues during the plant's life within the meristems of the stems and roots. Tobacco and soybean embryos contains only about 10 distinct cell and tissue types each, yet the mRNA populations contain roughly 20,000 different transcript classes. This is in contrast to animal development, in which a reduction in the number of genes expressed occurs as embryogenesis proceeds.

The Process of Embryogenesis

- Assymetric cleavage of the zygote results in the formation of an embryo with a suspensor and embryo proper.
- *Embryo proper*: Gives rise to embryo.
- *Suspensor*: Anchors embryo to surrounding embryo sac and ovule tissue; serves as a conduit for nutrients from maternal sporophyte into

developing proembryo. Suspensor senesces after the heart stage and is not a functional part of the mature seed.

- Embryonic organs and tissue-types differentiate during globular-heart transition phase.

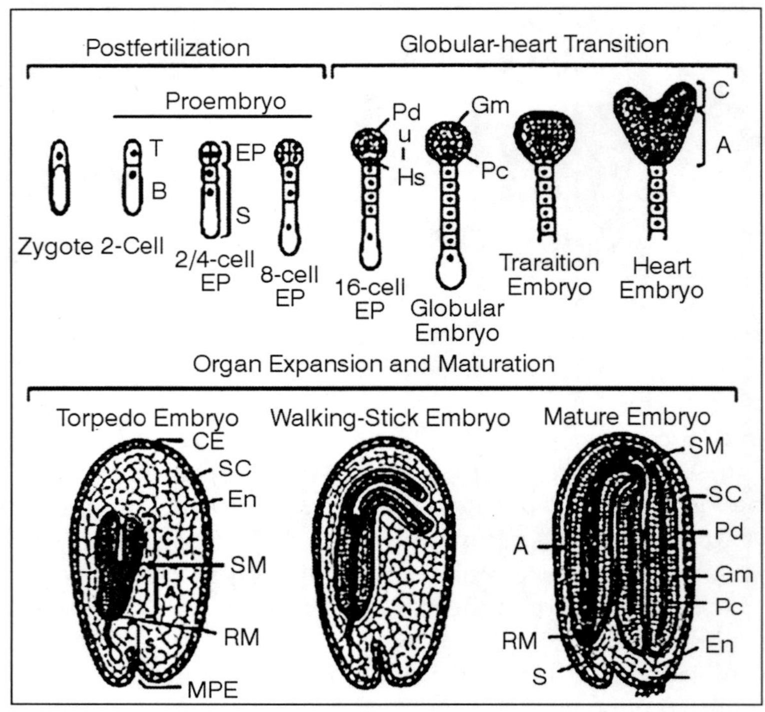

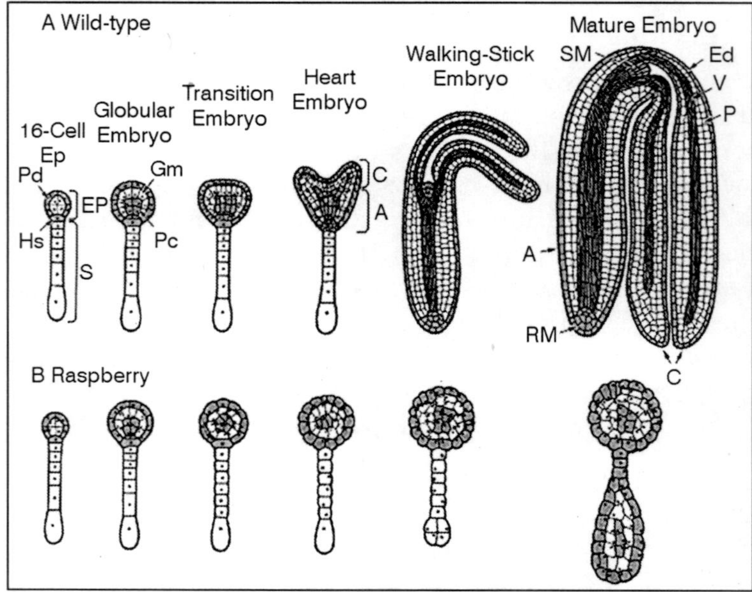

Fig. Overview of Embryogenesis

Most of embryo is devoted to formation of the root. The hypophysis (Hs) gives rise to the root meristem.

Hypophysis: The uppermost cell of the suspensor from which part of the root and rootcap in the embryo are derived.

The shoot further differentiates into the axis and cotyledon. After the critical heart embryo stage, the shoot meristem is determined. The shoot meristem will give rise to the epicotyl, which in turn, will mature into the bulk of the plant body.

DNA FRAGMENT AND CLONING VECTORS

A cloning vector is a small piece of DNA into which a foreign DNA fragment can be inserted. The insertion of the fragment into the cloning vector is carried out by treating the vehicle and the foreign DNA with the same restriction enzyme, then ligating the fragments together. There are many types of cloning vectors. Genetically engineered plasmids and bacteriophages (such as phage λ) are perhaps most commonly used for this purpose. Other types of cloning vectors include bacterial artificial chromosomes (BACs) and yeast artificial chromosomes (YACs).

Most commercial cloning vectors have key features that have made their use in molecular biology so widespread. In the case of expression vectors, the main purpose of these vehicles is the controlled expression of a particular gene inside a convenient host organism (eg. E. coli). Control of expression can be very important; it is usually desirable to insert the target DNA into a site that is under the control of a particular promoter. Some commonly used promoters are T7 promoters, lac promoters (bla promoter) and cauliflower mosaic virus's 35s promoter (for plant vectors).

To allow for convenient and favourable insertions, most cloning vectors have had nearly all their restriction sites engineered out of them and a synthetic multiple cloning site (MCS) inserted that contains many restriction sites. MCSs allow for insertions of DNA into the vector to be targeted and possibly directed in a chosen orientation. A selectable marker, such as an antibiotic resistance is often carried by the vector to allow the selection of positively transformed cells. All plasmids must carry a functional origin of replication.

Some other possible features present in cloning vectors are: vir genes for plant transformation, intergrase sites for chromosomal insertion, lacZα fragment for α complementation and blue-white selection, and/or reporter genes in frame with and flanking the MCS to facilitate the production of recombinant proteins [eg. fused to the Green fluorescent protein (GFP) or to the glutathione S-transferase.

TYPES OF VECTORS

- Plasmids.
- Bacteriophage.

- Cosmids.
- Yeast artificial chromosomes (YACs).
- Bacterial artificial chromosomes (BACs).

PLASMIDS

Characteristics of Plasmids

- Extrachromosomal circular DNA molecules which are not part of the bacterial genome.
- Size range: 1-200 kb.
- carry functions advantageous to the host such as.
  - Produce enzymes which degrade antibiotics or heavy metals
  - Produce restriction and modifying enzymes
- Replication is coupled to host replication in a:
  - Stringent manner: One (or two) plasmids made during each round of bacterial replication
  - Relaxed manner: 10-200 copies of the plasmid made during each round of bacterial replication; (this can be increased to 1000-2000 plasmids by stopping host protein synthesis and replication with the antibiotic chloramphenicol)

Foreign DNA is inserted into a plasmid (or any cloning vector) by ligating the DNA into a complementary site in the plasmid. These sites are generated by digesting the DNA and vector with the same restriction enzyme. (The site for the restriction enzyme that is chosen should only be represented once in the plasmid. Thus, when the plasmid is digested, a single, linear molecule would be generated.) The foreign DNA is then inserted into the plasmid by the action of the enzyme DNA ligase. The next step is to insert the ligated DNA into a bacterial cell for propagation. This is done by a technique called transformation. Bacterial cells are treated with either Ca2Cl or Rb2Cl. This treatment produces pores in the bacterial cell wall and membrane through which the plasmid enters. Although there is no size limitation to the ligation reaction, transformation efficiency is dictated by the size of the plasmid.

Process by which a plasmid is used to import recombinant DNA into a host cell for cloning. Many diseases are caused by gene alterations. Our understanding of genetic diseases was greatly increased by information gained from DNA cloning. In DNA cloning, a DNA fragment that contains a gene of interest is inserted into a cloning vector or plasmid.

The plasmid carrying genes for antibiotic resistance, and a DNA strand, which contains the gene of interest, are both cut with the same restriction endonuclease. The plasmid is opened up and the gene is freed from its parent DNA strand.

They have complementary "sticky ends." The opened plasmid and the freed gene are mixed with DNA ligase, which reforms the two pieces as recombinant DNA Plasmids + copies of the DNA fragment produce quantities of recombinant DNA. This recombinant DNA stew is allowed to transform a bacterial culture, which is then exposed to antibiotics. All the cells except those which have been encoded by the plasmid DNA recombinant are killed, leaving a cell culture containing the desired recombinant DNA. DNA cloning allows a copy of any specific part of a DNA (or RNA) sequence to be selected among many others and produced in an unlimited amount. This technique is the first stage of most of the genetic engineering experiments: production of DNA libraries, PCR, DNA sequencing, *et al.*

Inserting a DNA Sample into a Plasmid

Plasmids are similar to viruses, but lack a protein coat and cannot move from cell to cell in the same fashion as a virus. Plasmid vectors are small circular molecules of double stranded DNA derived from natural plasmids that occur in bacterial cells. A piece of DNA can be inserted into a plasmid if both the circular plasmid and the source of DNA have recognition sites for the same restriction endonuclease.

Restriction enzymes, also called restriction nucleases (EcoRI in this example), surrounds the DNA molecule at the point it seeks(sequence GAATTC). It cuts one strand of the DNA double helix at one point and the second strand at a different, complementary point (between the G and the A base). The separated pieces have single stranded "sticky-ends," which allow the complementary pieces to combine. The newly joined pieces are stabilized by DNA ligase. EcoRI, one of many restriction enzymes, is obtained from the bacteria Escherichia coli.

The plasmid and the foreign DNA are cut by this restriction endonuclease (EcoRI in this example) producing intermediates with sticky and complementary ends. Those two intermediates recombine by base-pairing and are linked by the action of DNA ligase. A new plasmid containing the foreign DNA as an insert is obtained. A few mismatches occur, producing an undesirable recombinant. The new plasmid can be introduced into bacterial cells that can produce many copies of the inserted DNA. This technique is called DNA cloning.

Effect of Plasmid Size on Transformation Efficiency

Molecule size (kb)	*% Maximum probability*
2.0	57
3.2	100
4.3	86
12.5	43
20.0	36
39.0	14
54.0	6

The goal of the ligation reaction is to insert the foreign DNA into the vector. An unwanted ligation product also occurs - religated vector. One way to minimize this event is to treat the vector with phosphatase. This removes the terminal phosphate group from the restriction site of the vector and theoretically prevents religation of the two ends. In reality though, this treatment is never 100%, so a low level of religation does occur. Thus after transformation you will have two types of cells: those which contain the original plasmid and those that contain a plasmid containing foreign DNA.

Plasmids are designed to distinguish the two types of transformation products. pBR322, the first widely used vector, utilizes differential antibiotic screening to distinguish the two types of transformation products. Let's say that we clone into the BamHI site of the vector. The insert DNA will then split the gene responsible for tetracycline resistance. But at the same time, the gene for ampicillin resistance is left intact. Transformed cells are first grown on bacterial plates containing ampicillin. This will kill all the cells that do not contain a plasmid. But we still cannot say which cells contain foreign DNA. Those cells that grew on ampicillin are then replica plated on plates with ampicillin and tetracycline. Those cells which grow in the presence of the ampicillin, but die under tetracycline selection contain plasmids which have foreign DNA inserts.

pBR322 was a breakthrough for molecular biology, but the double screening procedure was time consuming and could be subject to error. In 1981, a new series of plasmids were developed that permitted the identification of the foreign DNA containing cells in a single screening step. These are called the pUC plasmids. As with pBR322, ampicillin resistance is used as one selectable marker. The second marker is based on insertional inactivation of the E. coli lacZ gene. The wild type gene can hydrolyze a specific dye [X-Gal (5-bromo-4-indoyl-B-D- galactopyranoside)] to a blue colour, and the bacterial colony is stained blue. A multiple cloning site has been inserted into this gene. This site will accept fragments ending in a number of different restriction enzymes. Upon insertion of DNA into this site, the activity of the gene is eliminated and the colony appears white in colour. Thus transformed colonies containing plasmids with inserts can be distinguished from those with plasmids without inserts based on the colour of the colony and the ability of the colony to grow on an ampicillin containing media.

cDNA Cloning (Cloning Eukaryotic mRNA)

cDNA cloning is a method of obtaining a DNA copy of the mRNAs that are expressed at a specific stage in the development of the plant. In this manner, you can enrich the library that you will screen for those sequences in which you are interested. The reagent required for this type of cloning approach is mRNA. These mRNAs have a poly A+ tail. This tail permits the isolation of poly A+ mRNA by using either oligo-dT or oligo-U columns. Total

RNA is run through one of these columns under conditions which favour the binding of the tail to the matrix on the column. After the column is extensively washed, the conditions are changed and the bound mRNA is isolated. This is the starting reagent for cDNA cloning.

Steps in cDNA Cloning

- Bind oligo-dT to the poly +A tail of the mRNA.
- Add reverse transcriptase and make a DNA copy of mRNA (cDNA). Erase the mRNA with alkali and high temperature. (First strand synthesis)
- Add a C-tail to the 3' end of the cDNA with terminal transferase.
- Add oligo-dG to the tailed-cDNA and make the second strand with reverse transcriptase or the Klenow fragment of DNA Polymerase I.
- Add dC's to the 3' end of the double-stranded cDNA with terminal transferase
- Add C-tailed, ds-cDNA to G-tailed, PstI cut pBR322 and anneal. (DNA ligase is not required for this step.)
- Transform E. coli cells.
- Select Tetr/Amps cells. Most, but not all inserts can be removed by PstI digestions.

BACTERIOPHAGE LAMBDA VECTORS

Extensive research has been directed towards the development of multipurpose lambda vectors for cloning ever since the potential of using coliphage lambda as a cloning vector was recognized in the late 1970s. An understanding of the intrinsic molecular organization and of the genetic events which determine lysis or lysogeny in lambda has allowed investigators to modify it to suit the specific requirements of gene manipulations. Unwanted restriction sites have been altered and arranged together into suitable polylinkers. The development of a highly efficient in vitro packaging system has permitted the introduction of chimeric molecules into hosts. Biological containment of recombinants has been achieved by introducing amber mutations into the lambda genome and by using specific amber suppressor hosts. Taking advantage of the limited range of genome size (78 to 105% of the wild-type size) for its efficient packaging, an array of vectors has been devised to accommodate inserts of a wide size range, the limit being 24 kbp in Charon 40. The central dispensable fragment of the lambda genome can be replaced by a fragment of heterologous DNA, leading to the construction of replacement vectors such as Charon and EMBL. Alternatively, small DNA fragments can be inserted without removing the dispensable region of the lambda genome, as in lambda gt10 and lambda gt11 vectors. In addition, the introduction of many other desirable properties, such as NotI and SfiI sites in polylinkers (*e.g.*, lambda gt22), T7 and T3 promoters for the in vitro transcription (*e.g.*, lambda DASH), and the mechanism for in vivo excision of the intact insert (*e.g.*, lambda ZAP), has facilitated both cloning and subsequent analysis. In most cases, the recombinants can be differentiated from the parental phages by their altered phenotype. Libraries constructed in lambda vectors are screened easily with antibody or nucleic acid probes since several thousand clones can be plated on a single petri dish. Besides the availability of a wide range of lambda vectors, many related techniques such as rapid isolation of lambda DNA, a high efficiency of commercially available in vitro packaging extracts, and in vitro amplification of DNA via the polymerase chain reaction have collectively contributed to lambda's becoming one of the most powerful and popular tools for molecular cloning.

We have talked about plasmids as vectors for cloning small pieces of DNA. The limitation of this vector is the size of DNA that can be introduced into the cell by transformation. This presents problems when you are trying to create a genomic library of a large genome such as with plants. A genomic library contains all of the DNA found in the cell of the plant (or any organism). If you digest plant DNA to completion with a restriction enzyme, ligate those fragments into a plasmid vector and transform bacterial cells, only a portion of those fragments will be represented in the final transformation products. If a gene of interest is located on a large fragment then you will not be able to isolate that gene from a plasmid library. But what can be done to increase the probability of obtaining a clone which contains the entire gene. First you need to use a vector that can accept large fragments of DNA. Examples of these are bacteriophage and cosmid vectors and more recently yeast artificial chromosomes. Bacteriophage lambda vectors were developed because several observations were made that suggested that they could complete their life cycles even if foreign DNA was inserted into a portion of its genome. This suggested that certain regions of the virus were not essential. Let's first discuss the life cycle of lambda.

- *Adsorption*: The phage particle binds at a maltose receptor site of the bacterial cell; growing the cell in the presence of the sugar increase the number of receptor sites
- *Penetration*: DNA is injected into the cell; at this point it can enter one of two pathways;
  - *Lysogenic pathway*: The phage DNA becomes integrated into the genome and is replicated along with the bacterial DNA; it remains integrated until it enters the lytic pathway
  - *Lytic pathway*: Large scale production of bacteriophage particles that eventually leads to the lysis of the cell; base pairing at the cos site leads toa circular molecule.

The linear double-stranded DNA lambda genome contains about 50,000 nucleotide pairs and encodes 50-60 different proteins. When the lambda DNA enters the cell the ends join to form a circular DNA molecule. The bacteriophage can multiply in E. coli by a lytic pathway, which destroys the cell, or it can enter a latent prophage state. Damage to a cell carrying a lambda prophage induces the prophage to exit from the host chromosome and shift to lytic growth (green arrows). The entrance and exit of the lambda DNA from the bacterial chromosome are site-specific recombination events

- *Early transcription*: Transcription proceeds from the pL and pR promoters, through the N and cro genes and stops at terminators tL and tR1; a low level of transcription through the O and P genes occurs and terminates at tR2; the N product is an antitermination factor that is important for the next stage of transcription.
- *Delayed early transcription*: The N product binds to RNA polymerase and transcription proceeds past the tL, tR1 and tR2 terminators; genes to the left of N, involved in recombination, to the right of cro, involved in replication, are expressed at this point; another protein expressed from the Q gene is used for antitermination of later transcription.
- *Replication*: Early replication is through a theta form initiated from a single origin of replication site; later replication is via rolling circle replication; this produces long concatamers of the phage DNA that are cleaved at the cosL and cosR sites.
- *Late transcription*: The protein product of the cro gene builds up to a critical level and then binds to the oL and oR to stop early transcription; another protein, a product of the Q gene, has built up and activates transcription at the p'R promoter by antitermination; transcription terminates with in the b region; this transcription results in the production of the proteins required for the head and tail of the mature phage particle and those required for bacterial cell lysis.
- *Assembly*: A prophage head is produced; a unit length DNA is placed into the head by the action of the Nu1 and A proteins; the DNA is locked into place by the D protein and ter function of the A protein clips the DNA at the cosL and cosR sites; the concatamer is released, the tail is added and the mature phage particle is completed.

Packaging of the DNA into the head does not require a complete length of wild type lambda. It has been determined that a lambda molecule that is between 78% and 105% of wild type length can be packaged. This is from 37 to 53 kb in length.

Two important developments suggested that lambda may be suitable as a cloning vector. First it was determined that the gene products between the J and N genes could be removed and the life cycle could be completed. Second, restriction enzyme sites could be eliminated which permitted the development of a vector with a single site for insertion of foreign DNA.

Two types of vectors have been developed:

- *Insertional vector*: DNA is inserted into a specific site.
- *Replacement vector:* Foreign DNA replaces a piece of DNA (stuffer fragment) of the vector.

Let's talk about a specific vector EMBL 3 and EMBL 4. One important concern when cloning with lambda vectors is that you want to maximize the number of resulting phage particles that contain foreign DNA. Or said another way you want to minimize the number of wild type particles. One approach is through spi selection. This refers to sensitivity to P2 interference.

Table. Bacteriophage Phenotypes for Growth on P2 Lysogens

Phenotype	*Growth on P2 Lysogens? (bacterial strain)*
spi+ (red+gam+)	Poor
spi– (red–gam–)	Good

EMBL 3/4 vectors have placed the red and gam genes in the stuffer fragment. Thus only those particles from which the stuffer has been replaced can grow well in a P2 lysogen bacterial cell.

Cloning in Lambda Vectors

Phages are viruses that can infect bacteria. The major advantage of the phage vector is its high transformation efficiency, about 1000 times more efficient than the plasmid vector. The DNA to be cloned is first inserted into the DNA, replacing a nonessential region. Then, by an in vitro assembly system, the virion carrying the recombinant DNA can be formed. The genome is 49 kb in length which can carry up to 25 kb foreign DNA.

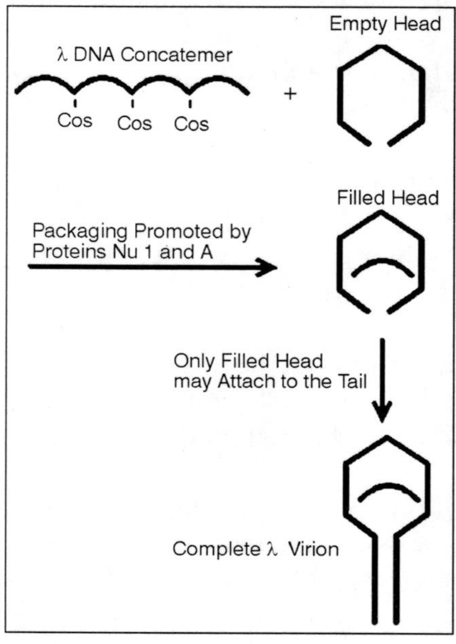

Fig. The Assembly Process of the Virion.

The extreme ends of the DNA are known as COS sites, each is single stranded, 12 nucleotides long. Because their sequences are complementary to each other, one end of DNA may base-pair with the other end of a different DNA, forming concatemers. The two ends of a DNA may also bind together, forming a circular DNA. In the host cell, the DNA circularizes because ligase may seal the join of the COS sites. In the assembly process of virions, two proteins Nu1 and A can recognize the COS site, directing the insertion of the DNA between them into an empty head.

The filled head is then attached to the tail, forming a complete virion. The whole process normally takes place in the host cell. However, to prepare the virion carrying recombinant DNA, the following in vitro assembly system is commonly used. Proteins Nu1 and A are encoded by the genes in the genome. If the two genes are mutated, DNA cannot be packaged into the pre-assembled head. Because tails attach only to filled heads, the cell will accumulate separate empty heads and tails, which can then be extracted. When the extract is mixed with recombinant DNA and proteins Nu1 and A, the complete virion carrying recombinant DNA will be assembled.

COSMIDS

Cosmids are plasmid vectors that contain cos sites. The cos site is the only requirement for DNA to be packaged into a phage particle. Cosmids were developed in light of this observation. How do you clone into cosmid vectors?

- Clone the DNA into the vector as you would with any plasmid.
- Introduce the DNA into the bacterial cell via a phage particle.
- Propagate as plasmid.

Since phage particles can accept between 38 and 53 kb of DNA and since most cosmids are about 5 kb, between 33 and 48 kb of DNA can cloned in these vectors.

It has the following advantages:

- High transformation efficiency.
- The cosmid vector can carry up to 45 kb whereas plasmid and phage vectors are limited to 25 kb.

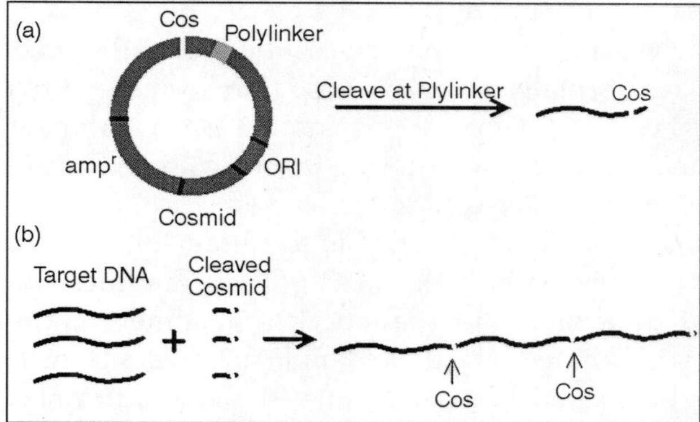

Fig. Cloning by using Cosmid Vectors.

(a) In Addition to Ampr, ORI, and Polylinker as in the Plasmid Vector, the Cosmid Vector also Contains a COS Site. (b) After Cosmid Vectors Are Cleaved with Restriction Enzyme, they are Ligated with DNA Fragments. The Subsequent Assembly and Transformation Steps are the same as Cloning with ? Phages.

Problems Associated with Lambda and Cosmid Cloning

- Since repeats occur in eukaryotic DNA rearrangements can occur via recombination of the repeats present on the DNA inserted into lambda or cosmid.
- Cosmids are difficult to maintain in a bacterial cell because they are somewhat unstable.

YEAST ARTIFICIAL CHROMOSOMES (YACS)

A yeast artificial chromosome (YAC) is a vector used to clone DNA fragments larger than 100 kb and up to 3000 kb. YACs are useful for the physical mapping of complex genomes and for the cloning of large genes. First described in 1983 by Murray and Szostak, a YAC is an artificially constructed chromosome that contains a centromere, telomeres and an autonomous replicating sequence (ARS) element, which are required for replication and preservation of YAC in yeast cells.

ARS elements are thought to act as replication origins. A YAC is built using an initial circular plasmid, which is typically broken into two linear molecules using restriction enzymes; DNA ligase is then used to ligate a sequence or gene of interest between the two linear molecules, forming a single large linear piece of DNA.

A plasmid-derived origin of replication (ori) and an antibiotic resistance gene allow the YAC vector to be amplified and selected for in E.coli. TRP1 and URA3 genes are included in the YAC vector to provide a selection system for identifying transformed yeast cells that include YAC by complementing recessive alleles trp1 and ura3 in yeast host cell. YAC vector cloning site for foreign DNA is located within the SUP4 gene. This gene compensates for a mutation in the yeast host cell that causes the accumulation of red pigment. The host cells are normally red, and those transformed with YAC only, will form colourless colonies. Cloning of a foreign DNA fragment into the YAC causes insertional inactivation, restoring the red colour. Therefore the colonies that contain the foreign DNA fragment are red.

One goal of molecular genetics is to obtain physical data about the genomic organization of long stretches of DNA. Traditionally, this data has been obtained by a technique called chromosome walking. Walking is performed by subcloning the end of a lamda or cosmid clone and screening a library for other clones that contain similar sequence information. If this new clone overlaps a portion of the original clone, then the length of the DNA of interest is extended by the length of DNA in the second clone that is not found in the original clone.

By performing these steps successive times, a long distance map can be obtained. This technique though has difficulties. First, each step is technically slow. Second, if you use lambda or cosmid clones, you might only extend the region of interest by 5-10 kb in each step of the walk. Finally, if any of the clones that are obtained contain repeated sequences, the subclone could lead you to another region of the genome that is not contiguous with the region of interest.

Yeast artificial chromosomes can alleviate some of these problems because of the large (100-1000kb) amount of DNA that can be cloned. First of all, YACs cannot speed up each step of the walk because the subcloning and screening steps can only be performed so quickly. But they can solve the other two problems.

Because they carry large amounts of DNA, each step can easily extend the region of interest by 50-100 kb and up to as much as 500 kb. Thus a long distance map of the region can be obtained in several steps. Secondly, although repetitive regions may be 10-20 kb in length they are rarely, longer than 50 kb. Thus a YAC with 100kb will contain some region that is single copy which can be used for further steps in the walk.

Features of YACs:

- Large DNA (>100 kb) is ligated between two arms. Each arm ends with a yeasttelomere so that the product can be stabilized in the yeast cell. Interestingly, larger YACs are more stable than shorter ones, which favours cloning of large stretches of DNA.
- One arm contains an autonomous replication sequence (ARS), a centromere (CEN) and aselectable marker (trp1). The other arm contains a second selectable marker (ura3).
- Insertion of DNA into the cloning site inactivates a mutant expressed in the vector DNA and red yeast colonies appear.
- Transformants are identified as those red colonies which grow in a yeast cell that is mutant for trp1 and ura3. This ensures that the cell has received an artificial chromosome with both telomeres (because of complementation of the two mutants) and the artificial chromosome contains insert DNA (because the cell is red).

DNA cloning is a technique that allows the wholesale production of a specific DNA sequence. DNA containing a gene of interest is inserted into the purified DNA genome of a self replicating element, which can be a plasmid, a virus or in this case, a yeast artificial chromosome (YAC). A YAC can be considered as a functional artificial chromosome (self replicating element), since it includes three specific DNA sequences that enable it to propagate from one cell to its offspring:

- *TEL*: The telomere which is located at each chromosome end, protects the linear DNA from degradation by nucleases.
- *CEN*: The centromere which is the attachment site for mitotic spindle fibers, "pulls" one copy of each duplicated chromo-some into each new daughter cell.
- *ORI*: Replication origin sequences which are specific DNA sequences that allow the DNA replication machinery to assemble on the DNA and move at the replication forks.

It also contains few other specific sequences like:

- *A and B*: selectable markers that allow the easy isolation of yeast cells that have taken up the artificial chromosome.
- Recognition site for the two restriction enzymes EcoRI and BamHI.

While DNA cloning into a plasmid allows the insertion of DNA fragment of about 10,000 nucleotide base pairs, DNA cloning into a YAC allows the insertion of DNA fragments up to 1,000,000 nucleotide base pairs. Why is it so important to be able to clone such large sequences? To map the entire human genome (3x1,000,000,000 nucleotide base pairs) it would require more than 100,000 plasmid clones. In principle, the human genome could be represented in about 10,000 YAC clones.

BACTERIAL ARTIFICIAL CHROMOSOMES (BACS)

A bacterial artificial chromosome (BAC) is a DNA construct, based on a functional fertility plasmid (or F-plasmid), used for transforming and cloning in bacteria, usually E. coli. F-plasmids play a crucial role because they contain partition genes that promote the even distribution of plasmids after bacterial cell division. The bacterial artificial chromosome's usual insert size is 150-350 kbp, but can be greater than 700 kbp. A similar cloning vector called a PAC has also been produced from the bacterial P1-plasmid.

Bacterial Artificial Chromosomes (BAC) have been developed to hold much larger pieces of DNA than a plasmid can. BAC vectors were originally created from part of an unusual plasmid present in some bacteria called the F' plasmid. The F' plasmid allows bacteria to have "sex" (well, sort of: F' helps bacteria give its genome to another bacteria but this only happens rarely when bacteria are under a lot of stress). F' had been studied extensively and it was found that it could hold up to a million basepairs of DNA from another bacteria. Also, F' has origins of replication and bacteria have a way to control how F' is copied. In 1992, Hiroaki Shizuya took the parts of F' that were important, cleaned it up, and turned it into a vector.

BAC vectors are able to hold up to 350 kb of DNA and have all of the tools that a vector needs to work properly, like replication origins, antibiotic resistance genes, and convenient places where clone DNA can insert itself. With these vectors it is possible to study larger genes, several genes at once, or entire viral genomes. By using a vector that can hold larger pieces of DNA, the number of clones required to cover the human genome six times theoretically could drop from 1.8 billion to about 50 million. Researchers have modified BAC vectors to become more convenient to use and more useful in specialized situations. In addition to the antibiotic resistance gene that was added to identify transfected bacteria, a gene was added that enabled the bacteria to turn the colourless substance X-gal/IPTG blue. This substance is found in the chicken-soup-like media on which the bacteria colonizes.

This colour-changing gene, called lacZ, is split apart when the clone DNA is incorporated into the vector, so it is possible to tell not only if a bacteria had been transfected (meaning incorporated into the cell), but also if the bacteria was transfected with the vector containing insert DNA or just the vector alone (remembering that if the vector has properly incorporated the clone DNA, it will have lost its ability to change X-gal/IPTG).

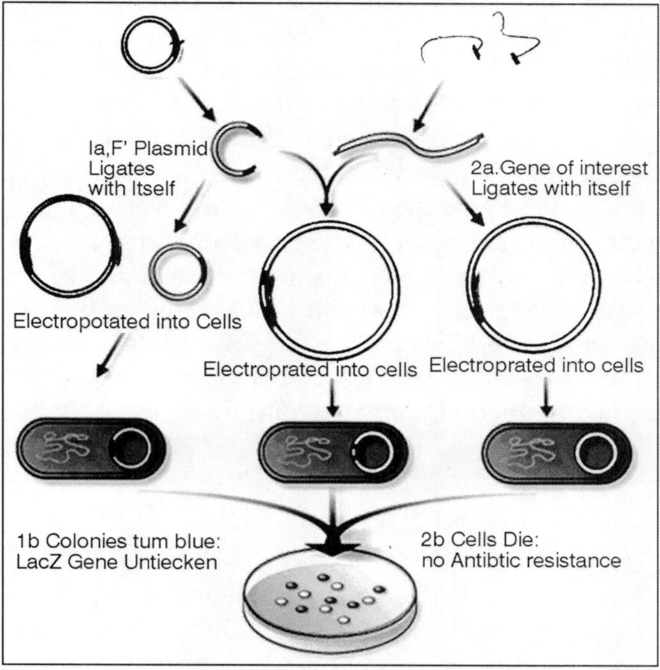

Fig. Selecting for Transformed Bacteria

Another similar modification uses the gene called sacB, which encodes a protein called levansucrase. This protein turns sucrose - table sugar - into levan, a toxic substance to bacteria. In a similar fashion, bacteria grown on media with sucrose will die if sacB is not broken up by inserted DNA. If the vector carries a DNA insert, then sacB is broken and it won't produce levansucrase, so the bacteria can survive in the sucrose media. In theory, only bacteria transfected with a vector containing insert DNA would be able to grow and form colonies.

Some modifications to BAC vectors make them more specialized. For instance, there are researchers studying the herpes virus who have made a

BAC vector that can be cultured in bacterial cells and then when put into a mammalian cell, instantly releases its insert DNA - in this case, the whole herpes virus genome. With such a vector, it is easier to grow sufficient amounts of the herpes virus for research, since it can live in bacterial cultures, instead of requiring their endemic mammalian cell cultures, which are extremely difficult to maintain. This in turn makes it easier to make modifications to the virus and study what each of its genes do. The creation of BAC vectors has allowed researchers to do many things that they could not do before and do them more quickly and more easily. Some of the scientific practices that have simplified with BAC vectors includes phylogenetic studies (studies which examine species' relationship to one another) and what the absolute minimum size of a genome could be.

The world is currently overrun by a plethora of microbial species that cannot be grown in cultures. BACs have allowed researchers to look at microbial DNA without having to actually grow the organisms, since the DNA is kept within easy-to-grow bacterial cultures. BAC vectors are also useful for studying pathogens, and are helpful in the development of vaccines. Many pathogens are becoming resistant to all antibiotics available to medicine and BAC vectors are playing a tremendous role in discovering new and powerful antibiotics in the environment. Discovering enzymes that are able to help clean up oil spills or help farmers breed healthier farm animals or even process radioactive waste are just a few examples of what Bacterial Artificial Chromosomes can do. BAC vectors are useful tools and the methods for working with them are fairly well developed. Continuing advances such the modification of experimental animals will only increase the wide variety of uses for big bad BACs.

LIBRARY SCREENING AND GENE SEQUENCING

Once a library is constructed it is screened for a particular gene of interest. Screening is based on homology between a probe and one of the clones in the library. The probe is normally a nucleic acid that has some sequence homology to the gene that is represented in the library. The library is the collection of clones from the source DNA which is inserted into a cloning vector. For example, a bean genomic lambda library contains pieces of the entire complement of bean DNA, from 9-20 kb in size, cloned into a lambda vector. Any probe used to screen the library should have some homology to a clone in the bean library, and this homology would allow you to select the appropriate clone from the many clones that do not contain your DNA of interest.

Probes come in several forms. The more homologous (similar in DNA sequence) the probe is to the sequence that is being sought, the easier it is to select a clone from the library. For example, a bean lambda library is considered a genomic library because it contains all the DNA sequences found in the bean genome. A bean leaf cDNA library, though, would contain only those sequences that are expressed in the leaf after they have undergone processing.

Thus the cDNA clones would not contain any of the intron sequences or the controlling elements of the gene. To obtain a genomic clone from the bean lambda library, the best probe would be a cDNA clone obtained from screening a bean cDNA library. This type of probe, a probe that contains the exact sequence of the sequence that is being sought, is called a homologous probe. But how was the original cDNA clone obtained so that it could be used as a probe.

The cDNA library from which the clone was obtained might have been screened by a probe from another species which represents coding information for the same gene but from another species. For example, a bean leaf cDNA library may be screened with a tomato RUBISCO small subunit clone. A probe that contains DNA which encodes for the gene of interest but from another species is called a heterologous probe. Many plant genes have been cloned by screening libraries with both homologous and heterologous probes.

Polymerase chain reaction techniques are now commonly used to clone genes. To use this technique primers must be designed that are complementary to your target sequence. One oligonucleotide will be complementary to the anticoding strand (the DNA strand of the gene complementary to the mRNA and used as a template for transcription). The second oligonucleotide will be complementary to the coding strand (the DNA strand of the gene complementary to the anticoding strand).

If the gene has been cloned in another species, you can use that sequence information to design primers to amplifiy a fragment of the gene from the DNA of your species. Often, the gene you have an interest in has never been cloned, but you may have isolated the protein. If this is the case, the first step of this method is to obtain partial protein sequence information of the protein. Micropeptide sequencers are available that can rapidly generate sequence information. From this sequence, you can use reverse translation of the amino acid sequence to obtain the nucleic acid sequences of an amino-terminal fragment and a carboxy-terminal fragment. In this case the fragment complementary to anticoding strand will be a direct conversion of the genetic code for the amino acids in the amino-terminal fragment of the protein. The strand complementary to the coding strand will be complementary to the derived nucleic acid sequence of the carboxy-terminal fragment. Let's say that the following is the N-terminal sequence of the peptide for which you want to derive a synthetic oligonucleotide:

NH_3+–Met–Cys–Val–Lys–Thr–Pro–CO_2–

The following would be an appropriate probe for the gene.

5'-A T G T G T/G G T N A A A/G A G N C C – 3'

(N = A, T, G, and C)

When you are constructing these probes several concepts should be kept in mind. First, the genetic code is written in the mRNA sequence. This synthetic oligonucleotide will thus, be complementary to the anit-coding strand that is used as the template for the mRNA. Secondly, because you only know the amino acid and not the DNA sequence you must deal with redundancies of the genetic code.

For example leucine can be represented by CTA, CTT, CTG, CTC. Thus your synthetic oligonucleotide needs to contain all the possibilities. Therefore the above sequence is actually a mixture of 64 (4×4×2×2) different sequences. Because of this redundancy it is always best to make you oligonucleotide one short of full length.

This will help some of the redundancy problems. In this example, the probe is 17 nucleotides long and the redundancies for proline are eliminated. Another method of reducing the redundancy problem is to incorporate the nucleotide deoxyinosine at any position where a high level or redundancy occurs. Deoxyinosine is a modified nucleotide that does not pair with any of the four bases found in DNA. Therefore, it does not affect the homology of the probe.

The following may be the amino acid sequence of a carboxy-terminal fragment:

NH_3^+–Phe–Tyr–His–Thr–Val–Ala–CO_2^-

The appropriate oligonucleotide sequence from the above sequence for PCR amplification would be:

5'- G C N A C N G T A/G T G A/G T A A/G A A -3'

Amplification of DNA between these two oligonucleotides will provide a homologous probe because it will represent the exact sequence of the gene for which you are searching. Once this probe is radiolabelled by nick-translation (as all probes must be radilabelled before library screening), it can be used to screen a cDNA or genomic library for a clone complementary to the amplified fragment.

DNA Sequencing With the Chain-Termination, Dideoxy Technique of Sanger

Once you have a candidate clone, you will want to sequence it to determine if it is similar to another gene already sequenced or whether it is unique. DNA sequencing is a DNA replication based reaction. The two requirements for DNA replication are a DNA template and a free 3'-OH group. These requirements must be met for any sequencing procedure. The following steps illustrate the Sanger procedure, the most widely used DNA sequencing procedure.

- Although this technique originally used DNA cloned in a vector that can generate single- stranded DNA, sequencing is now routinely performed in double-stranded plasmid or cosmid vectors. pUC plasmids and related plasmids are popular for double stranded sequencing. To begin the procedure, the DNA is made single stranded by a combination of heating and alkaline conditions.

- A primer which has a free 3'-OH group is annealed to the vector. A convenient primer is the multiple cloning site from the pUC plasmids. DNA synthesis will initiate here. 3. Four reactions are performed, each containing dATP, dGTP, TTP and 32P-dCTP and one dideoxy nucleotide in low concentration. Dideoxy nucleotides have a hydrogen rather than a - OH group attached to the #3 carbon and cannot be used for further extension of the chain. Thus in this reaction, various synthesis products will end with that nucleotide.
- Add DNA polymerase to the reaction. Nucleotides will be added one at a time but the chain will terminate when a dideoxy nucleotide is inserted. Because the dideoxy nucleotide is limiting a series of products, each ending with the same base, will be obtained.
- Separate the fragments on an acrylamide gel, develop via autoradiography and read the sequence.

7

Indian Agriculture and Biotechnology: The Challenge of the Next Millennium

The world has won the important battle in the area of food security, but the war is still on. A total of 800 million people, that is one out of every six persons, in the developing world do not have access to food. One-third of all pre-school children in the developing countries are food insecure. We, thus, have a big challenge ahead of us as we enter the next millennium. It is true that the mass starvation that was predicted for Asia in seventies and eighties did not occur. It is only because Science was effectively put to work to raise agricultural productivity. In India, the ' *Green Revolution'* was a success due to the introduction of improved seeds, fertilizer, irrigation and plant-protection measures combined with positive policy support, liberal public funding for agricultural research and development, and dedicated work of farmers. Notwithstanding all-round achievements, the basic problems of food security, poverty, equity and sustainability, continue to be a cause of concern in India today.

BIOTECHNOLOGY IN INDIA : A PROMISING FUTURE

After becoming an IT bellwether, India is now shifting its focus to the most promising industry of the future, Biotechnology. With its large pool of scientific talent, world-class informationtechnology industry, and vibrant pharmaceutical sector, India is well positioned to emerge as a significant player in the global biotech arena.

Biotechnology is perceived as a revolution throughout the world. Scientists, through Research and Development (R&D), have developed and are continuing to develop cures for diseases that have affected people for decades and even centuries. Scientists recently, have also, clinically developed crops that can withstand the brutalities of weather changes, helping poor farmers of the developing countries to retain their yield and increase their output manifold. Biotech, considered a boon by some, provides great hope to many around the world, and its benefits are and will be realized by more and more people over the years.

On the threshold of this new revolution, numerous companies have sprung up to take a piece of the exponentially growing Biotech market worldwide. The ever-decreasing physical boundaries enable biotech companies from the West to tap large markets around the world.

India to this extent holds a good advantage over many other countries of the world. With its large population of over a billion people there is a huge market for products and services. India's population has a very interesting demography that creates almost a perfect environment for biotech companies to shift bases here. In addition the Indian sub-continent, which occupies only 2.4% of the total global surface area, has the most varied species of flora and fauna. A study shows that, in percentage terms, India has about 7.6% of total mammal species, 12.6% of bird species, 11.7% of fishes and roughly 6.0% of total flowering plants that are present in the world. Biotech companies, by moving to India, can utilize this immense Bio-diversity, can easily find samples and, also conduct field research much more efficiently. Adding to this, India has one of the largest agriculture sectors in the world, and varied climatic zones that can help in research and development of different agribiotech products applicable worldwide

INDIAN ADVANTAGE IN AGRICULTURE

India, today, holds a small share of the global biotech market, but has all the capabilities to become a dominant player. The consumption of biotech products in India is expected to quadruple in the next decade. The human and animal segment of the industry alone is growing by at least 20%.

India has a rich human capital, which is the strongest asset for this knowledge-based industry. India has a large English speaking base and, according to Confederation of Indian Industry estimates, produces roughly 2.5 million graduates in IT, engineering and life sciences, about 650,000 postgraduates and nearly 1500 PhDs qualified in biosciences and engineering each year. India has proved its competency in selected areas of biotechnology such as, to name a few: capacity in bioprocess engineering, skills in gene manipulation of microbes and animal cells, capacity in downstream processing and isolation methods, and its competence in recombinant DNA technology of plants and animals. India has also allowed assisted stem cell research that permits researchers to use embryos from fertility clinics upon informed consent of the donors, thus giving it a clear head start in this new and promising field in Biotech. Clearly, India has the strength and capabilities in this industry, and a definite advantage to forge ahead and become the chosen location for many biotech companies looking for large markets and low cost qualified manpower to work in their R&D division.

INDIA'S BIO INDUSTRY: SEGMENT REVIEW

The Indian Biotech Industry can be divided into different segments. Following is a review of largest and the fastest growing segments of the Biotech Industry in India.

MEDICAL BIOTECH SEGMENT

The Indian pharmaceutical market is growing very rapidly. According to a study by Mckinsey, Indian Pharma industry is expected to grow to an innovation-led US $25 billion industry by 2010 with a market capitalization of almost US $150 billion from the current US $5 billion generic based drug industry. The vaccine market is expected to grow by roughly 20%.

AGRI BIOTECH SEGMENT

- India being the second largest food producer, offers a huge market for biotechnology products, especially agribiotech products.
- India has an excellent scientific infrastructure in agriculture, rich bio-diversity and skilled and low cost human-power.
- In a report by Ernst & Young it is expected that the Nutraceuticals market is roughly US $532-638 million presently and growing.
- With its 8000 kilometer of coastline including Andaman & Nicobar and Lakshwadeep islands, India has a rich aqua culture and its Marine resource development holds great potential.

SERVICES SEGMENT

With increasing number of pharmaceutical companies finding it difficult to conduct entire drug discovery process-in-house they are looking for ways to minimize costs. India has become a very attractive base as the cost of infrastructure is relatively lower compared to other nations.

Foreign companies also benefit from cheaper qualified workforce available in India.

India produces enough qualified graduates each year thus companies looking to expand their operations can easily do so without facing a shortage in labor.

It is estimated that vaccines, contract research, agriculture and human health sectors comprise as much as two thirds of the total market. It is further estimated that health care products would dominate the Indian biotech market, roughly 40% of the total market by the year 2010 followed by agriculture of about 30%.

It is also estimated that contract research and bioinformatics would pick up and account for as much as 25% of the biotech market.

An estimation by CII shows that the Agri-Biotech would see growth rates of as much as 60%, Diagnostic and Therapeutics of about 25% and Vaccines of about 15%. These figures clearly indicate the prospects of the Biotech Industry in India.

GOVERNMENT AND STATE INITIATIVES

The Government of India realized the potential and benefits of this industry at an early stage and formed the Department of Biotechnology in 1986 that has now become the central agency, responsible for policy, promotion of R&D and for international cooperation and manufacturing activities. Some of the initiatives taken by the Government of India are as follows:

The Government of India has been increasing its outlays to provide financial support to this industry. Government of India is also setting up a venture capital fund, to support small and medium enterprises.

Good regulatory framework has been set up for approval of GM crops and rDNA products. Recently, the Government of India decided to make changes to the Drugs & Cosmetics Act to make it more globally compatible.

Indian Patents Bill recently passed by the Parliament allowing 20-year patent term, inline with provisions made by WTO and TRIPS.

India has a sound and widely acknowledged framework of bio-safety guidelines to deal with evaluation, monitoring and release of genetically engineered organisms and there are more than 106 institutional bio-safety committees. The initiatives taken by both Central government and State governments have given a big boost to the Biotech industry in India. Foreign companies looking for new markets and to expand facilities to much more economical locations can find India ever more open and responsive to their needs.

STATE LEVEL INITIATIVES

Many state governments have also realized the benefits/importance of this industry. State governments namely Andhra Pradesh, Tamil Nadu, Karnataka and Maharashtra have begun to formulate their policies, develop R&D centers to encourage and nurture the Biotech Industry. Following is a brief outline of some of the initiatives taken by a few state governments:

Andhra Pradesh

- Government of Andhra Pradesh, in collaboration with the ICICI Limited, has set up a knowledge Park near Hyderabad.
- Development of 'Genome Valley' and Biotech Park with state-of-the art features.

- The Government has initiated many business friendly policies such as 'Single Window Clearance' mechanism geared to relieve the problem of red tape.

Tamil Nadu

- The Government of Tamil Nadu is facilitating in the setting up of the biotechnology enterprise zones (biovalleys) along the lines of Silicon Valley to exploit the bio resources of the state.
- Four state-of-the art biotech parks, a bioinformatics and genome centre will be established, each of which would be leveraging the bioresources of the agro-ecological zones of Tamil Nadu.

Maharashtra

- Has an excellent intellectual infrastructure. Through nearly 1000 institutions, it produces around 163,000 trained technical personnel each year. Some of the best Centers of excellence in India are present in Maharashtra.
- The government is also promoting biotech parks, R&D centers, and pilot plant facilities for underway contract research by putting equity stakes in such projects.

Karnataka

- The Karnataka government has announced a biotech policy to promote this sector and is setting up an institute for bioinformatics in Banglore.
- In addition the state government is also creating a biotechnology fund that will have inflows from the biotech companies. This could be used for incubation of new projects and promotion of the sector in the state.
- Karnataka has planned to launch India's first state sponsored biotechnology venture capital fund to boost their initiatives.

Himachal Pradesh

- Himachal Pradesh has prepared a blue print for promotion of biotechnology industries in the state which includes setting up of biotechnology parks, conservation and exploitation of bio-resources, intensification of R&D, and promoting biotechnology entrepreneurship through tax concessions and relaxed labor laws.
- It is also proposed to provide research based support to the private companies in form of providing for instance, access to a data base of bio-resources which is being developed along with separate entries of endangered medicinal plants.
- The State government has recently announced 100% tax holidays for all biotechnology products up to the year 2012.

INSTITUTIONAL SETUP ANDBUDGETARY ALLOCATIONS IN INDIA

India is one of the first few countries, among the developing countries, to have recognized the importance of biotechnology as a tool to advance growth of agricultural and health sectors as early as in 1980s. India's Sixth Five Year Plan (1980-85) was the first policy document to cover biotechnology development in the country.

The plan document proposed to strengthen and develop capabilities in areas such as immunology, genetics, communicable diseases, etc. In this context, referring to the Council of Scientific and Industrial Research (CSIR), the document suggested to ensure coordination on inter-institutional, inter-agency and on multi-disciplinary basis, full utilization of existing facilities and infrastructures in major areas including biotechnology. Programmes in the area of biotechnology included, tissue culture application for medicinal and economic plants, fermentation technology and enzyme engineering for chemicals, antibiotics and other medical product development; agricultural and forest residues and slaughterhouse wastes utilization and emerging areas like genetic engineering and molecular biology.

The existing national laboratories under the S& T agencies, such as Indian Council of Medical esearch (ICMR) and Council for Scientific and Industrial Research (CSIR) had initiated several esearch programmes to fulfill the above plan objectives. At the top, an apex official agency *viz.* ational Biotechnology Board (NBTB) was set up in 1982, to spearhead development of biotechnology. The NBTB was chaired by Member (Science) of the Indian Planning ommission and had representation of almost all the S&T agencies in the country *viz.* epartment of Science and Technology (DST), Council for Scientific and Industrial Research CSIR), Indian Council of Agricultural Research (ICAR), Indian Council for Medical Research ICMR), Department of Atomic Energy (DAE) and the University Grants Commission (UGC).

NBTB was formed with the specific purpose of the identification of priority areas and for volving a long-term plan for the country in biotechnology as well as to initiate and promote uch activities as conducive for further development of various areas in biotechnology. The BTB issued the " Long Term Plan in Biotechnology for India" in April 1983. This document pelt out priorities for biotechnology in India in view of the national objectives such as self ufficiency in food, clothing and housing, adequate health and hygiene, provision of adequate nergy and transportation, protection of environment, gainful employment, industrial growth and alance in international trade. Later in 1986, NBTB graduated to a full-fledged government epartment called Department of Biotechnology.

At present in India, there are six major agencies responsible for financing and supporting esearch in the realm of biotechnology apart from other sciences. They are Department of cience and Technology (DST), Department of Biotechnology (DBT), Council of Scientific and ndustrial Research (CSIR), Indian Council of Medical Research (ICMR), Indian Council of griculture Research (ICAR) and University Grants Commission (UGC), Department of cientific and Industrial Research (DSIR). DST, DBT and DSIR are part of Ministry of Science nd Technology while ICMR is with Ministry of Health, ICAR with Ministry of Agriculture and UGC with Ministry of Human Resource and Development. DSIR is the funding agency for CSIR and both of them independently fund biotechnology related research programmes.

Allocations for all of these agencies have gone up in the last decade. Out of this, DBT is the only agency completely devoted to R&D in biotechnology. It is very difficult to estimate the total allocations for this sector per se from other aforementioned agencies as in some cases the allocations are not separately marked as allocations for biotechnology. One faces this kind of constraint especially with those organisations, which are focusing on technological solutions and are not committed for X or Y nature of technology. Thus separately accounting for biotechnology becomes difficult. It would probably become possible only if a detailed survey at the institutional level is undertaken. A broad idea about total allocations by major ndian funding agencies to Science and technology related projects and not necessarily to biotechnology alone. In case of UGC it gives a broad idea not only about S&T related projects but also about other research streams.

DEPARTMENT OF BIOTECHNOLOGY (DBT)

There has been a significant increase in Government of India's outlays for biotechnology over the past decade. Since the time of establishment, in 1986, the allocation for the Department has increased manifold. The budgetary allocations have gone up from Rs. 404 million in 1987-88 to Rs. 1,138 million in 1997-98 and by 2001-2002 it became Rs 1,863 million. Though the current price allocation figures may not give a complete picture but the budgeted figure for 2002-03 shows a doubling of allocation to Rs. 2,356 million. In India the developmental allocations are generally made for five years under the National Five Year Plans. The Government has recently finalized the Tenth Five Year Plan (2002-2007). The Working Group on Science and technology for the Tenth Five Year Plan, constituted by the Planning Commission, has proposed an outlay of Rs. 20,750 million for the period 2002 to 2007. This marks a sharp increase of 234 per cent from the budgetary provisions made during the Ninth Plan period (1997-2002) which totaled at Rs. 6,215 million only. The Vision Statement for the Tenth Five Year Plan enumerates the proposal for human genome sequences, proteomics, structural biology and bioinformatics.

The DBT has taken special precaution to find crucial balance between different sections of society as far as technology absorption is concerned apart from promoting industrial development. It is supporting low-cost biotechnology adoption programmes for socially backward communities. The programmes include vermi-composting, use of organic manure, silk-worm rearing, mushroom cultivation, etc. Some training cum demonstration programme are also being supported for them. Efforts are also on for gender mainstreaming. The DBT has launched 11 projects for women in the areas of waste management, bio pesticides, bio fertilizers, floriculture, fish farming for poor women in the rural areas. At the Golden Jubilee Biotechnology Park for Women, industrial modules have been allotted to women entrepreneurs for setting up units in the above-mentioned areas. This Park was inaugurated in 1998 near Chennai the capital of a southern Indian state Tamil Nadu. The Park is located in Siruseri, adjacent to Information Technology Park in an area of about 20 acres. It would have some central facilities for the entrepreneurs for technology resourcing, training and marketing. This is a first unique project of a joint effort by the central and state government in the realm of biotechnology.

CHANGING ROLE OF DBT

In recent times, DBT has taken up a proactive role in promotion of industry. DBT proposed a single window application-processing cell as part of a new regulatory system for the domestic biotechnology sector. The move formed part of the recent recommendations on biotechnology sector made by the Confederation of Indian Industry (CII). Besides recommending setting up of a single window application-processing cell at DBT, CII had also suggested a fixed time frame of 150 days for clearing new biotech proposals.

In this regard, CII had recommended a process whereby a new application would be sent by the single window agency to the Review Committee on Genetic Manipulation (RCGM), which in turn would be required to submit a scientific evaluation report within 60 days of receiving the applications. This report will be submitted to the relevant approval committee, identified by the end product category. For example, in case of agricultural products, it would go to the Genetic Engineering Approval Committee (GEAC), in case of pharmaceutical products to the Drugs & Pharma Approval Committee (DPAC), and in case of food products to the Biotech Foods Approval Committee (BFAC). The GEAC/ DPAC/ BFAC would be required to accord approval or rejection within 90 days of receiving the evaluation report from RCGM. In case of rejection of the application, the applicants will also have the right to appeal to the concerned approval committee. CII had also suggested that any additional information required by RCGM for completeness of the application form would have to be called for within 30 days of receipt of the application. Besides DBT, the recommendations were submitted by CII to 14 other agencies, including seven ministries. Most of the recommendations submitted by CII have been accepted by DBT.

The Council of Scientific and Industrial Research was established in 1942. It is India's largest research and development organization. It has 40 laboratories and 80 field stations/extension centres spread over the length and breadth of the country. The total allocation for CSIR in the year 2000-01 was Rs. 9120 million, which is 13 per cent higher from 1999-2000 (Rs. 7940 million). CSIR's Centre for Cellular and Molecular Biology (CCMB) incubated India's first recombinant protein product from a private company, Shantha Biotech, a hepatitis B vaccine, and it has numerous industrial relationships, including a joint venture with Biological E and Amersham Pharmacia to build DNA micro-arrays.

INDIAN COUNCIL OF MEDICAL RESEARCH (ICMR)

Another major institution working in the area of biotechnology is the Indian Council for Medical Research (ICMR) under the Ministry of Health. It is the apex body in the country to promote, coordinate and formulate biomedical and health research. Central Government gives full maintenance grant to the Council, to research in communicable diseases, contraception, maternity and child health, nutrition, non-communicable and basic research.

The total allocation for ICMR from the Central Government (Ministry of Health) was Rs. 1470 million in 2000-01, which was 21 per cent higher over the allocation of the previous year, that is 1999-2000 (Rs. 1160 million). The Council is also engaged in research on tribal health, traditional medicine and publication and dissemination of information. In the year 2001 ICMR has launched a major programme in the field of genomics (vector, microbial, human) with the initial allocation of Rs. 510 million. One of the major areas of focus is the disease susceptibility gene identification, especially for, communicable diseases like leprosy, tuberculosis, noncommunicable diseases as rheumatic fever or genetic diseases as thalissimia established four centres for developing molecular medicine at All India Institute of Medical Sciences (AIIMS), New Delhi, Sanjay Gandhi Post Graduate Institute of Medical Sciences (SGPGIMS), Lucknow, Jawaharlal Nehru University (JNU), New Delhi and Post Graduate Institute of Medical Education and Research (PGIMER), Chandigarh. Apart from this, ICMR has also established six biomedical informatics units in different parts of the country. It has proposed an allocation of Rs. 1000 million for the Tenth Five Year Plan.

INDIAN COUNCIL FOR AGRICULTURE RESEARCH (ICAR)

In India, agricultural research is being spearheaded by the Indian Council of Agricultural Research (ICAR) under the Ministry of Agriculture. The Council is engaged in conducting research in the field of agriculture, soil and water conservation, animal husbandry, fisheries, dairying, forestry and also agricultural education. The allocation for ICAR from the Ministry of Agriculture was Rs. 13,990 million in 2000-01, which was Rs. 12,060 million in the previous year. It has several research laboratories all over the country conducting research in biotechnology, besides using traditional breeding techniques for different research projects. ICAR has established a National Research Centre on Plant Biotechnology (NRCPB) at the Indian Agricultural Research Institute (IARI), Pusa, New Delhi, which is fully dedicated to work on plant biotechnology. Annual expenditure over the projects at NRCPB is Rs. 150 million. Apart from this ICAR has also supported two research networks, combining ICAR and even non-ICAR research laboratories to work on crops like rice, cotton, brasica and brinzal. ICAR is also implementing a World Bank supported programme called National Agriculture Technology Programme (NATP) through which huge allocations have been made at different research laboratories for strengthening infrastructure for biotechnological research.

ICAR has also collaborated with DBT and Rockfeller Foundation to jointly launch a National Rice Biotechnology Network (NRBN) in 1988. This project helped in evolving a culture of collaboration among different institutions, which has led to publication of several international papers in established journals. This network put together plant breeders and molecular biologists on the one hand and also provoked interests of private sector in R&D. For instance, Mahyco and Rockfeller Foundation worked together on identifying the relevant genes for saving Indian rice from brown plant hoppers. In this collaborative effort, interaction was not limited to private entities only. It is between private and public sector organisations also. Several universities also have tie ups with private sector organisations. For instance, Tamil Nadu Agricultural University (TNAU) attempted joint trials with Monsanto on weed resistance in soil. TNAU was also partner in monitoring the Bt Cotton field trials of Mahyco-Monsanto alliance. TNAU is also holding in exploratory talks with Rasi seeds (it is also in alliance with Monsanto) to conduct and monitor field trials of Bt crops. Similarly, rice research work at National Research Centre for Plant Biotechnology has attracted business interest of companies like Nath Seeds Ltd. and JK Agri-genetics.

HUMAN RESOURCE DEVELOPMENT AND TRAINING

The National Biotechnology Board had launched an integrated short-term training programme way back in 1984, to cope up with growing demand for highly trained manpower. In the first phase (1984-85), 5 universities were selected for initiating M.Sc./M.Tech programme in this multi-disciplinary area. Subsequently, in 1985-86 and 1986-87, the DBT has added 8 universities/institutions for M.Sc/M.Tech/Post-doctoral teaching programmes. Subsequently, DBT was entrusted with the responsibility of evolving curriculum for biotechnology courses and meet the demand for human resources in the field of biotechnology. In 1986-87 a model system of post-graduate/post-doctoral teaching in biotechnology in 7 universities/institutions was launched Some of the specialised M.Sc. courses in marine and agricultural biotechnology were launched in 1988-89 at 3 universities. In 1992-93, DBT supported a five year Integrated Programme in biochemical engineering and biotechnology in Indian Institute of Technology, Delhi and a post-doctoral programme at Indian Agricultural Research Institute, New Delhi.

Now, DBT is supporting 20 M.Sc. courses in general biotechnology, 4 in agricultural biotechnology, one each in medical and marine biotechnology while couple of diploma courses in molecular and biochemical technology 14. The total intake of students in the various postgraduate courses supported by the DBT in the country is around 550 per year. As a part of restructuring of the post doctoral research and training programme, DBT has scraped the on going programme with different institutions and has given this responsibility to Indian Institute of Science (IISc), Bangalore. This is to ensure competitive attitude and quality output in the life sciences. It is being proposed that IISc would award up to 75 fellowships of two-year duration in different streams of biotechnology.

As part of a wider effort for capacity building in institutes of higher learning, full-fledged departments of biotechnology are being set up. The Indian Institute of Science, Bangalore, Indian Institute of Information Technology and Management, Gwalior; and select Regional Engineering Colleges are setting up departments of biotechnology. The All India Council for Technical Education (AICTE) has already approved B.Tech. programmes in biotechnology in eight engineering colleges and has since been advised to develop a model curriculum for undergraduate programmes. All the new departments will have undergraduate, post-graduate and doctoral programmes. Special funding will be provided for the purpose in the Tenth Five Year Plan. Apart from expanding teaching of biotechnology at higher educational institutions a separate module on biotechnology would also be integrated with the school curriculum. The Department of Biotechnology of Government of India will provide the necessary outline of this module so that the National Council of Education Research and Training (NCERT) and the Boards of School Education would be accordingly advised.

Indian University Grants Commission has come out with a scheme to promote higher centres of learning at one place and assist them as much as possible. In this regard, Delhi based Jawaharlal Nehru University (JNU) has been identified by the UGC as centre for excellence in the areas of genomics, genetics and biotechnology. The University has received funds to the tune of Rs 300 million and is planning to start a new integrated M.Sc./Ph.D programme in life sciences and made to upgrade equipment and library facilities. The new integrated programme in life sciences will reduce the time taken by a scholar to complete his Ph.D by at least within two years. Now the scholar will not be required to undergo a separate M.Phil programme. JNU is aiming at 10 seats for the integrated course and another 20-25 seats in the School of Life Sciences. The University has so far received 40 proposals for possible projects, which can be pursued by them in these fields in the future. Out of the funds that JNU has finally managed to get from the University Grants Commission on its selection as the University with potential for excellence, Rs. 100 million have been set aside for upgradation of facilities and equipments. The rest of the Rs. 200 million would help University to subscribe to some of the 8,000 online journals, both in the filed of Science and social sciences. Besides this, JNU also announced recruitment of more researchers. The tie-ups with industry are also likely to grow. For instance, researchers at the Centre for Biotechnology (CBT) at Jawaharlal Nehru University in New Delhi have been working for four years on a recombinant anthrax vaccine and soon would start Phase I clinical trials in collaboration with Panacea Biotech Ltd. (New Delhi).

PRIVATE SECTOR PARTICIPATION ANDROLE OF FINANCING INSTITUTIONS

In India, biotechnology industry has grown over the past few years at a very rapid pace to reach a nsizeable scale in terms of turnover. According to the available estimates, the size of India's market for biotechnology products could be between US $ 1.5 to US $ 2.5 billion depending upon, among other factors, how a biotechnology product is defined. Of this the agriculture sector market is valued between US $ 450 to US $ 500 million and diagnostic/vaccines market at US $ 150 to US $ 420 million.

AGRICULTURAL BIOTECHNOLOGY

Some of the private foundations such as M. S. Swaminathan Research Foundation (MSSRF), Chennai have taken important initiatives in terms of bridging gap between technology development, its commercialisation and ultimately its diffusion. One of the important project MSSRF launched in early 1990s was establishment of Biovillages in India and China. The Biovillage approach aims at covering principles of ecological sustainability and economic profitability with equity. This project actively promoted interaction between society, industry and R&D institutions. Some of the firms such as Indo-American Hybrid Seeds Company, Bangalore and R&D institution such as Tamil Nadu Agricultural University were the prominent partners. This project boosted the demand for biofertilisers in the Southern Indian Villages.

A SEED INDUSTRY

In the agriculture sector a large number of companies have taken up activities related to biotechnology. Leaving aside subsidiaries of TNCs in India, the agribiotech companies can be classified into three broad groups. One, the larger integrated seed companies which are expanding their R&D to cover biotechnology like Mahyco, Indo-American Hybrid Seed, etc. to develop their own transgenics. Second group is that of smaller companies which have not been active in research or product development but have started employing techniques such as tissue culture for their breeding programmes *e.g.* companies like Kastur Rangan and Adikeshevalu. The third group may cover highly specialized technology companies that undertake services for specified research, like contract research organisations. This is a relatively new concept in the agriculture R&D in India. Some of the companies like Avesthagen qualify in this group.

At this moment all Indian firms are under lot of pressure for technological improvement of their seeds. This pressure emanates largely from their consumers and the growing market penetration by multinational seed corporations. As a result these firms are now exploring the possibilities to embark on biotechnology related research. For instance, a large sized seed company, JK Agri – Genetics has set up a separate division for biotechnology research in those crops in which its owning a larger share in the hybrid seed market. However, on the other side of the spectrum one finds old hands in the field of biotechnology like Indo-American Hybrid Seeds, Bangalore. This was one of the premier companies, which entered in the scene way back in 1992-93 but is still struggling with identification of relevant gene sequences, high capital cost of R&D leading to resource crunch for research and then, on top of that, shortage of skilled manpower.

The company officials point out that despite of so many institutes and universities they are not getting relevant manpower for absorption in R&D units. They are continuing with this urge simply because a breakthrough in biotechnology may help them in retaining their market share in vegetable hybrid seeds as their competitors are gradually tying up with TNCs for accessing their vast pool of gene sequences for crop improvement. Their problem become further confounded from the growing number of relevant genes or gene sequences coming under patent ownership of TNCs.

However, some of the larger companies, which have readily gone for alliances with TNCs, are engaged in back crossing only. For example Maharashtra based Mahyco Seeds which has a tieup with Monsanto has developed transgenic cotton seeds through back-crossing with the genes borrowed from Monsanto for pest resistance. Some companies like Ankur and Rasi, which are not that big in size, have also alliances with Monsanto for similar endeavors. The combined market share of Monsanto through these three tie-ups (in cotton) comes closer to 20 per cent which is next to the public sector National Seeds Corporation (NSC) (45 per cent).

In fact, NSC is also concerned about its declining share over the years and growing concentration in the private sector. In India, almost 8.9 million-hectare of land area are under cotton cultivation with 2.86 million tonnes of cotton lint a year. Since, independence nearly 150 hybrid varieties of cotton have been released and hybrids account for 60 per cent of cotton cultivation. Similarly, small companies such as Bangalore based Kastur Rangan and Adikeshevalu have entered in upscale tissue culture research and related plant development. Actually the tuning point in India's history of seed industry development was the announcement of Seed Policy (1988). This policy statement and related institutional reforms have given a boost to the participation of private sector in the growth of the industry. The private companies have increased their share in the industry from 20 per cent in 1981 to 73 per cent in 1997 and 76 per cent in 2001. There are nearly 150 registered seed companies. The market size for seeds has grown from Rs. 10,000 million in 1994-95 to Rs. 22,000 million in 1998-99. The interesting point to be noted is that the concentration in favour of organized private sector has grown many folds. The share of public sector seed supply has declined from 40 per cent to 25 per cent while share of unorganized sector has declined from 25 to 15 per cent in the same time period. The share of organized seed sector has grown from 35 per cent to 60 per cent.

BIOFERTILISERS AND BIOPESTICIDES

In recent past, private sector participation in production of biofertilisers has grown at a very high pace. The production has gone up from 2.5 tonnes in 1992-93 to 10.38 tonnes in 1999-00. Accordingly, the number of firms engaged in biofertilisers has also gone up from 35 in 1992-93 to 95 in 1999-00. The Biofertilisers Statistics being regularly brought by the Fertiliser Association of India shows that the production of fertilisers is concentrated in Western Indian states. They are Gujarat (3 units), Madhya Pradesh (7 units), Maharashtra (24 units) and Rajasthan (3 units). However, some of the Southern states like Tamil Nadu have also encouraged private industry to set up biofertiliser units. In three years time more than 13 units have come up in the state. There are several types of biofertilisers being marketed in India. Some of the prominent ones are Rhizobium, Azatobacter and Azospirillum. The Indian Council for Agriculture Research and Department of Biotechnology have actively encouraged application of rDNA technology for better quality Rhizobium and Azatobacter.

In order to help industry, DBT has established certain repositories to keep micro organisms. In case of biofertilisers, the established repository for microbes is "National Facility for Rhizobium Culture Collections", Division of Microbiology, IARI, New Delhi. The others are National Centre for Conservation and Utilisation of Blue-Green Algae, IARI, New Delhi, Microbial Type Culture Collections, Institute of Microbial Technology, Chandigarh, National Facility for Marine Cyanobacteria, Bharatidasan University, Tiruchurapalli, Facility for Mycorrhizal Culture Collections, Tata Energy Research Institute, New Delhi. However, its important to mention here that the demand for biofertiliser suffers from three major factors: poor and uneven quality, short shelf life and small contribution of crop yield for biopesticides. DBT established a Biocontrol Network Programme in 1989. The major emphasis in this programme is to develop better formulations and cost effective commercially viable biopesticides including microbial pesticides, parasitoids and bacteria for use under IPM. The project has 80 R&D projects and total allocations of 1.8 million.

MEDICAL BIOTECHNOLOGY

The companies in medical biotechnology in India can be divided into three broad categories. One is that of small startup companies that have indigenously developed biotech products, *e.g.*, Shantha Biotech and Bharat Biotech.

Then there are large companies, which have started responding to biotechnology and have in fact incorporated biotechnology in their work plan, for instance, Dr. Reddy's Laboratory (DRL), Ranbaxy Laboratories and Wockhardt Ltd. The third group has start-ups, which are all set to emerge as contract research organisations (CROs). Largely their work comes from TNCs.

Then there are companies like Biocon India which may not fit well in this kind of classification as they have an established presence in the industrial biotechnology (the fermentation sector) and a growing presence in the pharmaceutical sector, so eventually encompany our first and second category. Biocon set its sights on biopharmaceuticals and using its capabilities in a wide range of fermentation technologies by 1995.

Two years on, Biocon established Helix – a wholly owned subsidiary to develop its biopharma operations, which began with a range of anti-cholestrol statin drugs. Another subsidiary of this company, Clinigene International, was set up to initiate longitudinal clinical studies in select disease segments. Thus there is a synergy-based expansion, Biocon is now recognised as the country's leading biotech conglomerate.

SHANTHA BIOTECH

This is one of the leading examples of our first category of firms – small start ups with their own biotech products. Shantha Biotech has the credit of developing India's first world class HepatitisB vaccine and making it available at one third of the prevailing market price of imported vaccines. This company has an active biotechnology programme since 1994. Now Pfizer Ltd., the major Pharma TNC, has obtained the first refusal rights from the Hyderabad based Shantha Biotechnics Pvt. Ltd. for exclusively marketing the products to be developed by the latter in future. Earlier, Pfizer had entered into a co-marketing agreement with Shantha Biotech for marketing the latter's recombinant DNA vaccine for Hepatitis-B. Shantha Biotech is currently in advanced stages of discussion with one European Pharma major and about three US based pharma companies for its research projects.

The company plans to research on protein purification, molecular cloning and expression of native and synthetic genes. Shantha Biotech will also be offering polyclonal and monoclonal antibody development and formulation of certain types of vaccines. Shantha Biotech has developed in-house expertise in recombinant DNA technology and are very strong in development of cell lines for development of recombinant products. The company has invested Rs. 100 million in the biotech facility with external funding (from Bank of Oman).

DR REDDY'S LABORATORY (DRL)

Some of the companies like DRL have grown in the recent past. It was established in 1984. DRL is a big pharmaceuticals company and now is setting up biotech production facilities as per the US FDA specifications. The company has also identified biogenerics as significant market area which is estimated at $14 billion. It has pegged its brand value at Rs. 9,160 million as against last year's value at Rs. 2,850 million.

The biotechnology facility includes setting up of three class 10,000 bulk recombinant protein production suites and new formulations facility as well. The biotechnology business would cover therapeutics, vaccines and diagnostics. DRL has a research alliance with Centre for Cellular and Molecular Biology (CCMB) Hyderabad. DRL has also established a research subsidiary in Atlanta called Reddy US Therapeutics Inc., as well as a contract research subsidiary that will focus on genomics.

RANBAXY LABORATORIES

Ranbaxy, India's largest Pharma company with sales of more than $500 million in the year 2000, also views innovation in biotechnology as key to its future. It is one of the oldest post-independence firm founded in 1968. The company has branched out from creating new formulations of existing drugs and has half a dozen molecules under development. Ranbaxy has collaborations with several U.S. and European companies to develop new formulations and technologies. For example, Ranbaxy and Vectura Ltd. (Bath, U.K.) announced in 2001 that the Indian company's Ranbaxy B.V. subsidiary (the Netherlands, Antilles,) will develop oral formulations using Vectura's controlled release drug delivery technology, with Ranbaxy providing clinical development, scaleup, manufacturing and marketing expertise. Ranbaxy has set a model in India in terms of drug development. The model suggests that develop a molecule upto the level a domestic company can afford to do generally upto the first phase of the clinical trial and then outsource it to a leading MNC for further development and later on explore the possibilities for marketing tie-ups.

WOCKHARDT LTD.

Wockhardt Ltd. established nearly four decade ago, is the fifth largest pharmaceutical company of India. Wockhardt has grown at over 20 per cent annually for the last 5 years. In the year 2000, it acquired Merind from the

Tata's and besides Wallis Laboratories, a company in the UK, apart from entering into marketing alliance with the American Sidmak Labs. The company is formulating its biotechnology strategy around these initiatives. It has decided to split its business into two separate entities. Wockhardt Ltd. will remain as a pharmaceutical company while the new entity would focus on Life sciences only.

The new company (the name of which has yet to be finalized) will have all the other businesses namely hospitals, nutrients, IV fluid and agrovet (Crop protection). The R&D activity is also being categorized into three divisions. The first division concentrates on developing new bulk drugs (Novel drug delivery system (NDDS)) and generics. The second division concentrates on recombinant biotechnology. At present, the second division is focusing on technology absorption with the assistance of scientists from abroad. The third division is dedicated for discovery of new molecules. Wockhardt is expected to come out with its first investigational new drug (IND) in anti-infective therapeutic segment by end of 2002.

AGRICULTURE IN INDIA

India's record of progress in agriculture over the past four decades has been quite impressive. The agriculture sector has been successful in keeping pace with rising demand for food. The contribution of increased land area under agricultural production has declined over time and increases in production in the past two decades have been almost entirely due to increased productivity.

Contribution of agricultural growth to overall progress has been widespread. Increased productivity has helped to feed the poor, enhanced farm income and provided opportunities for both direct and indirect employment. The success of India's agriculture is attributed to a series of steps that led to availability of farm technologies which brought about dramatic increases in productivity in 70s and 80s often described as the Green Revolution era. The major sources of agricultural growth during this period were the spread of modern crop varieties, intensification of input use and investments leading to expansion in the irrigated area. In areas where 'Green Revolution' technologies had major impact, growth has now slowed.

New technologies are needed to push out yield frontiers, utilize inputs more efficiently and diversify to more sustainable and higher value cropping patterns. At the same time there is urgency to better exploit potential of rainfed and other less endowed areas if we are to meet targets of agricultural growth and poverty alleviation. Given the wide range of agroecological setting and producers, Indian agriculture is faced with a great diversity of needs, opportunities and prospects. Future growth needs to be more rapid, more widely distributed and better targeted. These challenges have profound implications for the way farmers' problems are conceived, researched and transferred to the farmers. On the one hand agricultural research will increasingly be required to address location specific problems facing the communities on the other the systems will have to position themselves in an increasingly competitive environment to generate and adopt cutting edge technologies to bear upon the solutions facing a vast majority of resource poor farmers.

In the past agriculture has played and will continue to play a dominant role in the growth of Indian economy in the foreseeable future. It represents the largest sector producing around 28 percent of the GDP, is the largest employer providing more than 60 percent of the jobs and is the prime arbiter of living standards for seventy percent of India's population living in the rural areas.

These factors together with a strong determination to achieve self-sufficiency in food grains production have ensured a high priority for agriculture sector in the successive development plans of the country.

An important facet of progress in agriculture is its success in eradication of its critical dependence on imported foodgrains. In the 1950's nearly 5 percent of the total foodgrains available in the country were imported. This dependence worsened during the 1960's when two severe drought years led to a sharp increase in import of foodgrains. During 1966 India had to import more than 10 million tonnes of foodgrains as against a domestic production of 72 million tonnes. In the following year again, nearly twelve million tonnes had to be imported. On the average well over seven percent of the total availability of foodgrains during the 1960s had to be imported.

Indian agriculture has progressed a long way from an era of frequent droughts and vulnerability to food shortages to becoming a significant exporter of agricultural commodities. This has been possible due to persistent efforts at harnessing the potential of land and water resources for agricultural purposes. Indian agriculture, which grew at the rate of about 1 percent per annum during the fifty years before independence, has grown at the rate of about 3 percent per annum in the post independence era.

AGRICULTURE – SUB-SECTORS

Indian agriculture broadly consists of four sub-sectors. Agriculture proper including all food-crops oilseeds, fiber, plantation crops, fruits and vegetables is the largest accounting for nearly 70 percent of the agriculture sector as a whole. The rapid growth in this sub-sector through exploitation of wastelands and fallows, spread of irrigation and adoption of production enhancing technologies was critical in transforming India from a country vulnerable to food shortages to one of exportable surplus. Although this sub-sector has made impressive progress its share in the sector as a whole has declined from 78 percent in 1960-61 to less than 70 percent by early 90s.

Table. Agriculture sub-sectors – value of output (%)

Sub-sector	1960-61	1970-71	1980-81	1993-94
Agriculture	78.0	79.3	75.3	68.5
Livestock	16.6	15.1	17.2	24.5
Forestry	4.4	4.4	5.9	3.8
Fisheries	1.0	1.2	1.6	3.2
	100.0	100.0	100.0	100.0
Total value at Current price (Rs. Billion)	89	218	615	2740

Correspondingly the share of livestock sector has increased considerably. The livestock industry has grown from Rs. 15 billion in early 1960s to Rs. 100 billion by 1980-81 and Rs. 672 billion by 1993-94. In nominal terms the sector grew at almost 15 percent per annum during 1980s. Milk production, which was almost stagnant for two decades ending 1970, grew by over 5 percent per annum in the 80s. Similarly, production of eggs increased at the rate of about 6.5 percent during the same period. As a result the share of livestock increased from about 17 percent till early 80s to 25 percent by 1993-94.

Though it plays relatively a minor role within the sector as a whole, fishing sub-sector activities have been on the rise. The sub-sector has grown from only Rs. 3 billion in 1970-71 to nearly Rs. 90 billion in 1993-94. The growth was particularly rapid in 70s and 80s. Value added increased at over 5 percent per annum during this period.

In real terms forestry and logging activities have been on the decline since mid seventies. As of 1993-94, the size of the industry in terms of value of output was 103 billion.

Table. Foodgrains Production

Year	*Area (million ha)*	*Production (million tonnes)*	*Yield kg/ha*
1950-51	97.32	50.82	522
1960-61	115.58	82.02	709
1970-71	124.32	108.42	872
1980-81	126.67	129.59	1023
1990-91	127.84	176.39	1379
1998-99	124.0	195.25	1574

Over the past three decades, the country has successfully transformed itself from a food deficit economy to one which is essentially self sufficient in availability of foodgrains and other essential commodities, albeit only at the prevailing level of effective demand. Annual aggregate foodgrains production, which averaged about 82 million tonnes in 1960-61 increased to 123.7 and 172.5 million tonnes for the trienniums ending 1980-81 and 1990-91 respectively. Current (1998-99) production level is 195 million tonnes and the country has been able to accumulate substantial, (35 million tonnes) stocks of foodgrains to cope up with any sudden difficulties arising from drought or a similar situation in any part of the country.

Increased outputs, have been achieved chiefly by adopting, since mid sixties, a strategy aimed at increasing foodgrains production by concentrating public sector efforts and resources in regions with a high potential for quick and substantial productivity gains through increased cropping intensity and average yields. These were the areas favoured by agroclimatic resource conditions and where irrigation facilities already existed or could be developed relatively rapidly. The main elements of this strategy were:

(i) expansion of irrigation coverage,
(ii) increased provision and utilization of key inputs – mainly high yielding varieties (HYVs) of crops, mainly of wheat and rice and chemical fertilizers and plant protection chemicals,
(iii) expansion and improvement of institutional support services such as research and extension and
(iv) price policies favourable to producers of major foodgrains.

The success of this strategy was made possible by development and availability of replicable production technology packages, so called 'Green Revolution' technologies. Irrigation facilitated double cropping and widespread adoption of HYVs. The HYVs performed particularly well under irrigated conditions, were highly responsive to fertilizers and their short duration permitted increases in cropping intensities.

Irrigation development was the cornerstone of the strategy. Undivided India was amongst the largest irrigated areas in the world. With partition nearly one-third of the irrigated area went to Pakistan. At the time of independence the net irrigated area was 20.9 million ha (gross irrigated area 22.6 million ha). Recognizing large-scale development of irrigation facilities as critical to rapid agricultural growth, the country has spent about Rs. 45,000 crores on irrigation development in the first four decades after independence. During the period 1950-51 to 1965-66 development of irrigation through government canals grew from 7.2 million ha to 9.8 million ha – a growth rate of 2.1 percent per annum. During 1970s this pace dropped slightly to 1.9 percent. In 1980s the rate of increase dropped significantly to 1.1 percent per annum.

The growth of tube-well irrigation, however, increased rapidly from 4.5 million ha in 1970-71 to 9.5 million ha in 1980-81 and then to 14.3 million ha by 1990-91. The net irrigated area increased from 31 million ha in 1970-71 to 53.5 million ha in 1995-96 which corresponds to 22 percent of the net sown area in 1970-71 and 37.63 percent in 1995-96 *(Tables 3, 4)*. With improvements in irrigation efficiency the gross irrigated areas has increased to 71.51 million ha. The percentage of gross cropped area service by irrigation increased from

18.3 percent in 1960-61 to 23.0 percent in 1970-71 and to over 38 percent at present.

Table. India – irrigated area

Year	*Net*	*Gross*	*Net as % of net sown*	*Gross as % of gross cropped*
	million ha		*%*	
1970-71	31.10	38.19	22.17	23.04
1980-81	38.72	49.78	27.66	28.83
1990-91	48.02	63.20	33.61	34.03
1995-96	53.51	71.51	37.63	38.33

Table. Sources of irrigation

Year	*Net Irrigated Area (million ha)*	*Sources*				
		Canals	*Tanks*	*tube-wells*	*Other wells*	*Other Sources*
		(% net irrigated area)				
1960-61	24.80	10.4	4.6	0.2	7.2	2.4
1970-71	31.10	41.28	13.22	14.34	23.88	7.29
1980-81	38.72	39.49	8.22	24.62	21.08	6.59
1990-91	48.02	36.34	6.13	29.69	21.73	6.11
1995-96	53.01	32.04	5.81	33.52	22.16	6.47

Fertilizers have constituted yet another key input in addition to expanded irrigation and spread of HYVs in achieving goals of high production and productivity. India currently occupies third position in the world, after China and USA, in terms of fertilizer production and consumption. Consumption of fertilizers has increased from 1.54 million tonnes in 1967-68, representing the pre green revolution era, to 17.31 million tonnes currently (1997-98).

The average per hectare use of fertilizers currently around 85 kg per hectares is the lowest among several Asian countries.

However, rice and wheat account for a major fraction, around 65 percent of the total fertilizer consumed in the country, with very little fertilizers going to the rainfed areas.

According to some current projections, fertilizer's use will need to increase to 30-35 million tonnes to meet the foodgrains need of 2020. The demand for nutrients will stretch by almost another 15 million tonnes if requirements for horticulture, vegetables and plantation and commercial crops are included. At present domestic production of N and P fertilizers (13.42 million tonnes) falls short of consumption by over 20 percent. In addition the entire requirement of K fertilizer is imported.

Most agricultural development programmes initiated in 1960s were concentrated in regions of high potential. Thus five states, Punjab, Haryana, Uttar Pradesh, Andhra Pradesh and Tamil Nadu account for 50 percent of the country's net irrigated and 53 percent of the gross irrigated area. The combination of expanding irrigation coverage and widespread adoption of short duration HYVs led to significant increases in cropping intensities. Acreage cropped more than once per year increased from 13 million ha in 1950-51 to about 44 million ha at present.

Average cropping intensity for the country as a whole rose from 115 percent in 1960-61 to 131 in 1993-94. By 1993-94 cropping intensity has risen to 187 percent in Punjab, 167 percent in Haryana and 142 percent in Uttar Pradesh.

Table. Cropping intensity

Years	*Net sown area*	*Gross cropped area*	*Area cropped more than once*	*Cropping intensity*
	million ha			*(%)*
1950-51	118.8	131.9	13.1	111.1
1960-61	133.2	152.8	19.6	114.3
1970-71	140.3	165.8	25.5	118.2
1980-81	140.0	172.6	32.6	123.3
1990-91	143.0	185.7	42.7	129.9
1993-94	142.0	186.4	44.4	131.2

An important consequence of the strategies adopted since sixties has been to boost production of, chiefly, two crops rice and wheat. Their share in total foodgrains production went up from 57 percent in 1970-71 to more than 75 percent in 1990-91. Production of foodgrains other than rice and wheat did not increase significantly and in the eastern region even the yield of rice did not increase.

Agricultural production and income rose substantially in the north-western states of Punjab, Haryana, Western Uttar Pradesh, parts of Rajesthan, Tamil Nadu and Andhra Pradesh. By contrast productivity and output growth have been modest in eastern and central India and in deccan plateau. Progress was particularly slow in rainfed areas, which account for over 60 percent of the cropped area and where a great majority of rural poor are concentrated. An important impact of the strategies pursued in the 'Green Revolution' period has been intensification of regional disparities and imbalances in agricultural development and food availability and hence levels of food security. For the country as a whole while per capita availability of cereals has increased substantially, that of pulses has decreased significantly.

In summary an annual increase in foodgrains production of 3.22 percent during fifties was mainly because of expansion in area. Sixties recorded a low annual growth rate of 1.72 percent necessitating large-scale imports of foodgrains.

Table. Rice and Wheat Production
Production (million tonnes)

Year	*Rice*	*Wheat*	*Foodgrains*	*R+W as % of Foodgrains*
	million tonnes			
1970-71	40.80	20.86	108.4	57.4
1980-81	49.91	34.55	129.6	65.4
1990-91	72.78	53.03	176.4	75.0
1996-97	81.74	69.35	199.4	75.0

Annual growth of 2.08 percent was recorded during seventies. This decade was the turning point in India's foodgrains economy leading to self-sufficiency through significant productivity increase first in wheat and later in rice in the eighties. An annual growth of 3.5 percent in foodgrains in eighties was the hallmark of green revolution that enabled India to become self-sufficient and even a marginal exporter. The pace of growth slowed in nineties barely making or even slower than the population growth rate. This is a matter of concern.

Table. Annual compound growth rate of foodgrains production

Crop	**1950-51 to 1959-60**	**1960-61 to 1969-70**	**1970-71 to 1979-80**	**1980-81 to 1989-90**	**1990-91 to 1997-98**
Rice	3.28	- 8.05	1.91	4.29	1.53
Wheat	4.51	5.90	4.69	4.24	3.67
Coarse cereals	2.75	1.48	0.74	0.74	- 0.49
Total cereals	3.00	2.51	2.37	3.63	1.84
Pulses	2.72	1.35	- 0.54	2.78	0.76
Total foodgrains	3.22	1.72	2.08	3.54	1.66

Table. Per capita net availability of cereals and pulses per day (g)

Year	*Cereals*	*Pulses*	*Total*
1951	334	61	395
1961	400	69	469
1971	418	51	469
1981	417	38	455
1991	468	42	510
1998	451	33	484

HORTICULTURE

The diversity of physiographic, climate and soil characteristics enables India to grow a large variety of horticultural crops – fruits, vegetables, flowers, spices, aromatic and medicinal plants, plantation crops etc. India is the largest producer of fruits in the world and second largest producer of vegetables. The area under fruits estimated at 1.45 million ha in 1970-71 grew to 2.8 million ha in 1991-92 and then more rapidly to 5 to 6 million ha by 1994-95. This sector is likely to grow rapidly in the future both on account of internal demands and export opportunities.

ANIMAL HUSBANDRY AND FISHERIES

Animal husbandry and dairying sub-sector plays an important role in overall economy and in social development. The contribution of the sub-sector is estimated to be about 25 percent of the total value of output of agricultural sector.

Table. Production of milk, eggs and fish in India

Year	*Milk production / availability*		*Fish*	*Eggs*
	million tonnes/annum	*g./day*	*('000 tonnes)*	*(million)*
1950-51	17.0	124	752	1832
1960-61	20.0	124	1160	2881
1970-71	22.0	112	1756	6172
1980-81	31.6	128	2442	10060
1990-91	53.9	176	3836	21101
1998-99	75.0	210	5388	28400

The sector also plays a significant role in supplementing family incomes and generating employment in the rural sector particularly among the landless, small and marginal farmers and women besides providing nutritious food. Production of suitable cross breeds and their wider adoptions has contributed to increasing country's milk production, which has now reached 75 million tonnes annually. Similarly, genetic improvement and better management practices have markedly pushed up production of poultry and eggs.

Table. Number of holdings (million)

Category of Farmers	*1970-71*	*1980-81*	*1990-91*
Marginal	35.7 (50.6)*	50.1 (56.4)	62.1 (59.0)
Small	13.4 (19.0)	16.1 (18.1)	20.0 (19.0)
Semi-medium	10.7 (15.2)	12.5 (14.0)	13.9 (13.2)
Medium	7.9 (11.2)	8.1 (9.1)	7.6 (7.3)
Large	2.8 (3.9)	2.2 (2.4)	1.7 (1.6)
All groups	70.5 (100)	89.0 (100)	105.3 (100)

*Figures in parenthesis represent the percentage of holdings.

Through the overall contribution of fisheries sub-sector is small, development of prolific and fast growing forms of several common fish species coupled with breakthrough in breeding under captivity, fish seed production and multi-layer fish culture has resulted in registering a very high annual growth of 11 percent in aquaculture production during the past decade.

Table. Area operated - % of total area

Category of Farmers	*1970-71*	*1980-81*	*1990-91*
Marginal	9.0	12.1	14.9
Small	11.9	14.2	17.3
Semi-medium	18.5	21.1	23.1
Medium	29.7	29.6	27.2
Large	30.9	23.0	17.4
All groups (million ha)	**162.1 (100)**	**163.8 (100)**	**165.6 (100)**

SUSTAINABILITY CONCERNS

Several indicators highlight increasing concerns of sustainability in areas which have largely contributed to increased production in the 'Green Revolution' era. Adoption of high yielding cultivators is virtually complete. Almost entire wheat and rice crops in the states of Punjab, Haryana and Western Uttar Pradesh are irrigated. In the higher production regions yields are plateauing and most traditional sources of productivity growth having been exhausted future gains in production have to come from elsewhere.

At farmers' level concerns are being expressed in several ways. Many farmers believe that the input levels have to be continuously increased in order to maintain high yields. In sixties and seventies most farmers used only nitrogenous and phosphate fertilizers to achieve high yields. Due to widespread deficiencies of several secondary and micronutrients, most farmers now have to apply higher doses and a greater variety of fertilizers to maintain crop yields.

Results from many long term studies on rice-wheat cropping system show a declining yield trend when input levels were kept constant – thus the growth rate of system productivity has been declining relative to growth rate of nutrients use.

Lowering of groundwater tables due to intensive rice-wheat system in many areas is resulting in increased costs of lifting water in the intensively cultivated high production areas, diseases and pest problems are turning more serious than ever before and pose both short and long large problems. It is reported that some weeds have developed resistance to the commonly used herbicides. What this implies is that the farmers are applying increasing amount of herbicide incurring increasing cost without the benefit of effective control. Pesticide residues entering the food chain and overall safety in use of pesticides continue to be serious problems.

Other emerging problems threatening sustainability of intensive cropping system *e.g.* rice-wheat include loss in biodiversity related issues. Large areas planted to a single/few varieties of a crop is a potential cause of concern. As the diversity is reduced natural processes that control and affect habitat quality and genetic expression weaken and for this reason internal and natural control mechanisms must be replaced by more externally applied artificial controls in the form of management and inputs which in due course lead the system towards unsustainability.

Groundwater is the major source of meeting the irrigation needs of irrigated agriculture. Currently about half the area under irrigation in the country is irrigated from groundwater sources.

Large-scale groundwater development has led to fall in the water table in many areas. Over pumping is leading to declining water table levels and failure of tube-wells. Pumping costs are increasing, as is the energy consumption. In the coastal areas this has led to ingress of sea water, with serious environmental implications.

Changes in water quality are adversely affecting agriculture and vice-versa. Inefficient and/or over use of fertilizers and pesticides in agriculture and untreated disposal of industrial and urban wastes are leading to increasing contamination by such elements as lead, zinc, copper, chromium, cadmium particularly in areas having high industrial activity *e.g.* in districts of Ludhiana, Faridabad, Kanpur, Varanasi etc.

An increase in the content of arsenic has been reported in several of the districts of West Bengal. This is attributed amongst other causes to the lowering of groundwater table due to excessive groundwater withdrawal and is leading to serious and widespread toxicity problems adversely affect the health of hundreds of thousands people of the region.

CHANGING LAND-USE AND FUTURE OF AGRICULTURE

One of the most important consequences of growing pressure on land is the declining trend in the average farm size and the pattern of holdings. According to the latest Agricultural Census in 1970-71 there were 70 million holdings operating 162 million ha.

By 1990-91 there were 105 million holdings operating 165 million ha. The average farm size decreased from 2.30 ha in 1970-71 to 1.57 ha in 1990-91. As of 1990-91 about 78 percent of holdings were small (1.0 to 2.0 ha) and marginal (<1.0 ha). A little more than 20 percent of the farmers were semi-medium (2.0 to 4.0 ha) and medium (4.0 to 10.0 ha).

Large farmers (>10.0 ha) constituted only 1.6 percent of the total holdings. Over the twenty-year period since 1970 the proportion of marginal farmer has increased from 50 to 59 percent and that of large farmer has declined from about 4 to 1.6 percent.

Table. India – Land Use Categories (million ha)

Classification	*1960-61*	*1970-71*	*1980-81*	*1990-91*	*1993-94*
Reporting area	298.5	305.0	305.0	305.0	305.0
Area under non-agricultural uses	14.8	16.5	19.7	21.1	22.0
Barren and un-cultivable	35.9	28.2	20.0	19.4	19.0
Cultivable waste	19.2	17.5	16.7	15.0	14.5
Old fallow	11.2	8.8	9.9	9.7	9.7
Forests	54.1	63.9	67.5	67.8	68.4
Permanent pastures	14.0	13.3	12.0	11.8	11.2
Land under misc. trees etc.	4.5	4.3	3.6	3.8	3.7
Net sown area	133.2	140.3	140.0	143.0	142.1
Current fallow	11.6	11.1	14.8	13.7	14.3

The proportion of total area operated by marginal farmers increased from nine percent in 1970-71 to nearly 15 percent in 1990-91 while the proportion of large farmers declined from about 31 percent to 17 percent in the same period. The size of average holding is very unevenly distributed among the states. States with relatively large average size of operational holding are a mixed lot – they include states with large tracts of barren lands *e.g.* Rajasthan, Maharashtra and Madhya Pradesh on the one hand and agriculturally advanced state like Punjab on the other. These trends in farm size changes will have a profound effect on the future agricultural development strategies.

Table. Livestock and poultry population (million)

Category	*1982*	*1987*	*1992*
Cattle	192.45	199.69	204.53
Buffaloes	69.78	75.96	83.49
Sheep	48.76	45.70	50.79
Goats	95.25	110.20	115.28
Total (Livestock)	419.59	445.28	470.15
Total Poultry	207.74	275.32	307.10

Although India's population growth rate has slowed from 2.1 percent in 1980s to 1.8 percent in the 1990s and is expected to slow further in the coming decades, yet the population is projected to reach 1.33 billion by 2020 from the current one billion. The urban share of total population is projected to increase from 26 percent to 35 percent of the total population. Although incidence of poverty is falling, it is estimated that in 93-94 (upto which data is available) 320 million people constituting 36 percent of the population were below the officially defined poverty line.

The nature of the poverty line has been shifting. About 30 years ago 48.4 percent of those living in rural areas were poor and 20 percent of those living in the urban areas were classed as poor. Recent studies show that the number of poor in urban areas have been increasing at relatively higher rate compared to the rural areas. At present those below the poverty line in rural sector constitute 37 percent of the population while in the urban sector the percentage is 32 percent. In the context of poverty alleviation, therefore, emphasis will be required to be placed both on production of food by the poor as well as on the availability of food for the urban poor.

It needs to be recognized that a large proportion of the rural poor are located in regions of low potential for food production *e.g.* arid and semi-arid areas, hilly regions, degraded land and forest areas. Widespread hunger and malnutrition are the direct manifestation of poverty and will call for increasing efforts to produce more food at affordable price.

Increasing population and economic growth are changing patterns of land use making potentially unsustainable demands on the country's natural resources.

Since early fifties the net area sown was expanded rapidly at first but at a diminishing rate since 1970 to reach approximately 142 million ha at present.

During 1950s and 1960s areas under agriculture expanded substantially as the fallows were reduced and cultivable wastes were put under the plough.

The net area sown increased from 119 million ha in 1950-51 to 133 million ha by 1960-61 and further to 140 million ha by 1970-71. Fallow lands declined from 28 million ha in 1950-51 to 20 million ha by 1970-71. Cultivable wastelands declined from 23 to 17.5 million ha.

Land use intensity *i.e.* fraction of net sown area to total geographical area increase from 36 percent in 1950-51 to 40.5 percent in 1960-61 and 43 percent by 1970-71 where it has since stabilized.

- Cropping intensity *i.e.* gross sown area as percent of net sown area increased from 111 percent in 1950-51 to 115 percent in 1960-61, 118 percent in 1970—71 and 130 percent by mid 1990s.
- While the contribution of increased area in the growth of agriculture has declined over time, that of productivity has increased. The yield of all crops grew at 1.5 percent per annum between early 1950s and mid 1960s. the pace accelerated to 1.7 percent in the 1970s and then to 3 percent per annum between early 1980s and mid 90s. Unlike the gains in area, which benefited non-foodgrains, the gains in productivity accrued mostly to foodgrains.
- India's forest resources have been dwindling. According to the 'State of Forest Report' (1997) the total forest cover of the country is estimated at 63.34 million ha *i.e.* 19.27 percent of the geographic area of the country. Of these the dense forest (crown density more than forty percent) and open forest (crown density 10 to 40 percent)

occupying about 11 and 8 percent of the geographic area respectively and mangroves occupy 0.15 percent of the geographic area. The country has lost about 5482 sq. km. of forest cover since the 1995 assessment. By any estimate the area under forest is far below the national policy goals and many areas nominally under forest are being used for non-forest purposes. Similarly 'uncultivated lands' such as permanent pastures, miscellaneous tree crops, cultivable wastes and fallow is subject to increasing competition from uses other than feeding livestock.

- The growth of livestock population is an important source of competition for land. The increase in number of major classes of livestock.
- The area sown to fodder crops is not recorded. Information available from other sources provide an estimate ranging from 4 to 5.5 percent of the net sown area and suggest that the area under fodder crops will have to increase to 10 percent or more to support increasing livestock based activity.

The pressure on India's land and water resources is seriously threatening native plant and animal diversity. India has uniquely rich and diverse genetic base. With increasing agriculture and economic development the genetic pool is declining. This decline, if unchecked and poorly managed can have unforeseen and adverse consequences for the sustainability of agriculture of the region.

AGRICULTURE IN THE CHANGING GLOBAL SCENARIO

Steady globalization of trade has profound implications for future agricultural development. The diversity of India's agro-ecological setting, high bio-diversity and relatively low cost of labour provide potential for agricultural competitiveness in a globalized economy. It is expected that with increasing globalization of markets over the years there will be demands for agricultural intensification. This will also be favoured because of greater backward and forward linkages between agriculture and food industry.

Therefore, increase in production and productivity are bound to be strategically important to economy. Intensification will not only favour alleviation of rural poverty but will also improve resource conservation particularly in the small farming sector where farmers can be encouraged to take up organized production of high value crops such as fruits, specialty vegetables, flowers medicinal and aromatic herbs etc. Stronger demands for crops of the small farmers' will not only improve incomes and welfare but will also make investments in technology and resource conservation more attractive. The General Agreement on Tariff and Trade (GATT) and liberalization of global trade is bound to have impact on future land use and production pattern. Understanding the local, national and international environment under which agricultural production is taking shape will be crucial in developing our own strategies.

EXTENSION STRATEGIES

Since early fifties a number of public by funded agricultural development programmes have been sponsored. These have included programmes like the National Extension Service (NES) Blocks in 1953, the Intensive Agricultural District Programme (IADP) in 1961-62, the Intensive Agricultural Area Programme (IAAP) 1964-65, the High Yielding Variety (HYV) programme 1966-67 and the Small and Marginal Farmers' Development Programmes (SMFDP) in 1969-70. Though these programmes had a perceptible impact the efforts did not get replicated over different areas and categories of farmers. In mid seventies based on pilot level project in Rajasthan Canal and Chambal command area a 'Training and Visit' (T&V) system of extension was promoted in different states. Extension efforts of the Indian Council of Agricultural Research through its research Institutes and the State Agricultural University were largely limited to demonstration of new technologies through such programmes as National Demonstration Project, Operational Research Project, the Lab to Land Programme and the Krishi Vigyan Kendras. However, there appears much to be desired in the way that extension programmes are conceived and implemented. At present extension programmes are implemented in largely a top-down fashion leaving little scope for localized planning and action. Farmers are almost passive receivers and their involvement in the process of technology generation and adoption is almost absent. Extension services, at present, are almost exclusively in the public sector domain and there is no effort or institutional support for other operators *e.g.* the NGOs, the corporate bodies etc.

Extension programmes sponsored by the government operate largely in isolation and there appears a strong need to view the extension programmes as an integral part of the research and development process.

The challenges facing agricultural development call for fundamental changes in our approach to technology transfer/extension programmes. Changes are necessary in the context of changing economic environment following policy adjustments in relation to privatization, deregulation and globalization calling for greater efficiency and effectiveness of the extension system. More importantly there is need for

- Greater emphasis on providing producers with knowledge and understanding needed to overcome the problems or to exploit opportunities of their own specific production systems. Correspondingly there will be a need to de-emphasize 'package of practices' or the blanket recommendations, top down approach followed thus far.

- Shift in the focus of public extension systems from promoting inputs use to one on sustainable management of resources and improvements in the production system as a whole.
- Closer interaction between farmers, extension scientists and production system researchers in diagnosing problems and identifying location specific recommendations emphasizing participation and education rather than being prescriptive.
- Widening the range of extension delivering agencies. While the publicly operated extension systems will continue to be important, there will appear a greater role for NGOs, farmers' associations and corporate sectors in particular situations. Role of commercial suppliers of seeds, agrochemicals, machinery, vaccines and medicines in providing advisories, as is already being done in a limited way, will need to be encouraged and factored into public system's own priorities.
- Wider and more creative use of mass media in tune with current developments in information technology to get information across to the farming community whose ability to overcome constrains at farm level will increasingly depend on access to reliable and up-to-date information.

TECHNOLOGICAL NEEDS AND FUTURE AGRICULTURE

It is apparent that the tasks of meeting the consumption needs of the projected population are going to be more difficult given the higher productivity base than in 1960s. There is also a growing realization that previous strategies of generating and promoting technologies have contributed to serious and widespread problems of environmental and natural resource degradation. This implies that in future the technologies that are developed and promoted must result not only in increased productivity level but also ensure that the quality of natural resource base is preserved and enhanced. In short, they lead to sustainable improvements in agricultural production.

Productivity gains during the 'Green Revolution' era were largely confined to relatively well endowed areas. Given the wide range of agroecological setting and producers, Indian agriculture is faced with a great diversity of needs, opportunities and prospects. Future growth needs to be more rapid, more widely distributed and better targeted. Responding to these challenges will call for more efficient and sustainable use of increasingly scarce land water and germplasm resources.

Technical solutions required to solve problems will be increasingly location-specific and matched to the huge agroecological/climatic diversity. Detailed indigenous knowledge and greater skills in blending modern and traditional technologies to enhance productive efficiency will be more than ever before, key to the farming success and sectoral growth. Most technological solutions will have to be generated and adapted locally to make them compatible with socio-economic conditions of farming community.

New technologies are needed to push the yield frontiers further, utilize inputs more efficiently and diversify to more sustainable and higher value cropping patterns. These are all knowledge intensive technologies that require both a strong research and extension system and skilled farmers but also a reinvigorated interface where the emphasis is on mutual exchange of information bringing advantages to all. At the same time potential of less favoured areas must be better exploited to meet the targets of growth and poverty alleviation.These challenges have profound implications for products of agricultural research. The way they are transferred to the farmers and indeed the way research is organized and conducted. One thing is, however, clear – the new generation of technologies will have to be much more site specific, based on high quality science and a heightened opportunity for end user participation in the identification of targets. These must be not only aimed at increasing farmers' technical knowledge and understanding of science based agriculture but also taking advantage of opportunities for full integration with indigenous knowledge. It will also need to take on the challenges of incorporating the socio-economic context and role of markets.

With the passage of time and accelerated by macro-economic reforms undertaken in recent years, the Institutional arrangements as well as the mode of functions of bodies responsible for providing technical underpinning to agricultural growth are proving increasingly inadequate. Changes are needed urgently to respond to new demands for agricultural technologies from several directions. Increasing pressure to maintain and enhance the integrity of degrading natural resources, changes in demands and opportunities arising from economic liberalization, unprecedented opportunities arising from advances in biotechnology, information revolution and most importantly the need and urgency to reach the poor and disadvantaged who have been by passed by the green revolution technologies.

Another important implication of increasing globalization relates to the need for greater attention to the quality of produce and products both for the domestic and the foreign markets. This would imply that production must be tuned to actual rapidly changing product demand. Such adaptation to global markets would require state of the art research, which can be achieved only by setting global standards of research, focus on well defined priorities and mechanisms which permit close interaction of farmers with researchers, the private sector and markets.

CHALLENGE OF INDIAN AGRICULTURE

There are several concerns about Indian agriculture that we need to address. Even in the green revolution crops, rice and wheat, the Indian average yield ranks around 50th in the world. A weak agro-based industry, the problems of weak marketing, storage, transportation, credit support, etc, haunt us. Only 0.3% of agricultural GDP is spent on agricultural R &D. The population is growing at an alarming rate of 1.8% per annum. It will be around 1.3 billion by 2020 and would require 50% additional foodgrains. A total of 200 million people are still below the poverty line and have insufficient access to food. Declining resource base, degeneration of natural resources, including soil fertility, water environment, etc, are some other concerns.

Food production systems centre around a few states only. Out of 17 big states, only 7 are self-sufficient and surplus in foodgrains. There was a significant decline in per capita intake of calories and protein between 1975 and 1990.

Inspite of all these problems, there is plenty of good news. India harbours a vast diversity of basic natural resources. Tremendous bio-diversity of agriculturally important micro-organisms exists, which are not yet been harnessed. We have a variety of crops, including quality fruits, vegetables, foodgrains, and a number of plantation and commercial crops. Indian fisheries resource has spread over 8,100 km of coastline on the east and west, and provides access to marine resources and transportation.

Very large proportion of area of various food crops is under low productivity category. It can be increased with inputs of high class science. There is a scope to cultivate sizeable portion of waste lands through soil amendment and introduction of agroforestry. Several million hectares of saturated soil can be potentially exploited from rainfed lowlands of eastern India.

The India Exclusive Economic Zone (EEZ) of the marine sector, of about 2 million sq.km., has high potential harvestable yields. Only about 20% of the cultivated area is covered under tractor cultivation, great scope of making quantism quantum jump in production and productivity exists, changing consumption patterns offer scope for diversified agriculture. Trade liberalization provides opportunities to reach for outside markets, which were not accessible otherwise. Great gains are possible through agro-based industrialization and overall rural development.

The cultivators and farmers of India have a rich heritage of traditional agricultural wisdom on crop and livestock husbandry and fisheries; based on the principles of the conservation of natural resources and environment. India has an extensive network of extension channels for dissemination of technologies generated by research institutes. India has a strong manufacturing capability and a large network of about 20,000 manufacturers and nearly one million village artisans for indigeneous production of all types of agricultural machinery. Therefore, inspite of several disadvantages, we have a great chance to leap forward, if we enforce the right policy initiatives and implement them.

BIOTECHNOLOGY AND INDIA AGRICULTURE

We need to recongnise that, just as the food reequirements of today's population of nearly 6 billion people could not have been met by the technologies of the 1940's, we cannot assume that current practices will feed the population of 8 billion expected by 2020. New approaches are needed in addition to the continued improvement of existing methods of crop and animal husbandry and food processing.

The strategic integration of biotechnology tools into India agricultural systems can revolutionise Indian farming and usher in a new era. Genetic engineering is clearly the most revolutionary tool to impact agricultural research since the discovery of genetics by Mendel. Prior to genetic engineering, the exchange of DNA material was possible only between individual organisms of the same species. With the advent of genetic engineering in 1972, scientists have been able to identify specific genes associated with desirable traits in one organism. For example, a gene from bacteria, virus or animals may be transferred into plants to produce genetically modified plants having changed characteristics. This method therefore allows mixing of the genetic material from species that cannot otherwise breed naturally.

Genetic engineering can be vital for an agrarian country like India. It can help in minimising the crop damage through disease and pest resistant varieties, reducing the use of chemicals, enhancing stress tolerance in crop plants, thus permitting productive farming on unproductive lands, etc. One could even extend the growing season of crops and minimise losses due to environmental factors on the one hand and increase the shelf life of fruits and vegetables, on the other, thus minimising losses due to food spoilage. We can expand the market vista and improve food quality. Biotechnology can also produce plants that possess healthy fats and oils, possess increased nutritive value, and create a whole range of higher value feeds. Biotechnology has even the power of producing biodegradable plastics, edible vaccines, etc.

It is only through the blending of the 'gene revolution' with our experience in the 'geen revolution' that we can reach our goal of 'evergreen revolution' and also 'nutritional revolution' The advantage of the gene revolution is that it is relatively scale neutral, benefiting big and small farmers alike, It is also environment friendly. Thus, it can be of great help to the smallest farmer with limited resources in increasing farm productivity through the availability of improved but powerful seed. It can also reduce a farmer's dependency on chemical inputs such as pesticides and fertilisers.

Modern biotechnology offers unlimited opportunity for enhancing genetic potential of crops and other commodities, management of biotic and abiotic stresses, bio-remediation and organic recycling. India is one of the main centres of agricultural biodiversity and its gene richness can greatly complement the developments in modern biotechnology. Likewise, new developments in GIS, remote sensing, and crop modelling, provide new opportunities for integrated management of natural resources. The revolution in informatics provides opportunities for sharing latest information for research and planning in a highly organised and efficient manner. With the globalisation of economy, the opportunities for value addition and post harvest management are also immense. With clever blending of technologies, the India farmer can usher in new era of confidence and performance. Sir Francis Bacon had once said: *'It would be an unsound fancy to expect that things which have never yet been done, can be done except by methods which have never been tried'* In the new context, we have to review the new role of biotechnology. Also, it is in this context that we have to view the rapid advances that modern biotechnology is making, including the area of transgenics. Let us focus on some of the developments in this area, since considerable interest and controversy has been created around the world today in this exciting area.

MOVEMENT IN TRANSGENICS

The advent of genetic modification over the last two decades has enabled plant breeders to develop new varieties of crops at a faster rate than was possible using traditional methods, with huge potential for further beneficial development.

Tobacco was the first plant to be genetically transformed\, in 1983, with cereals beginning in 1990, but it is only recently that products such as GM soya have reached the market place in Europe, and, for a variety of reasons, given rise to concern & controversy.

The concern must be addressed seriously if society is to exploit the new technologies appropriately. We will address this issue rather extensively a little later. Whereas in India, we have been discussing and debating the issues, the rest of the world is galloping ahead.

Table : Ag Biotech Product Sales

$ Millions		*Sales*		*Annual Growth Rate*	
	1994	1997	2002	1994-97	1997-02
Transgenic seeds and plants	20	405	2,030	173%	38%
Animal growth horomes	80	225	405	31	12
Biopesticides	35	65	110	23	11
Other	107	180	340	19	14
Total U.S. Sales	$242	$875	$2,885	53%	27%

Development, field testing, and commercialisation of transgenic crops, is now recongnized by the international scientific community to be an essential strategy for food security. Fortyfive countries, including India, have conducted transgenic crop field trials up to 1997. The activity in the developing countries has been shown in Table below.

Table : Releases of Transgenic Plants in Developing Countries (by Species and Introduced Trait)

Species	*Introduced Trait*					
	Total Fied releases	*Herbicide Resistance*	*Insect Resistance*	*Virus Resistance*	*Product quality*	*Others*
Maize	46	29	16	1	3	5
Soya bean	27	25	—	1	1	—
Cotton	24	16	15	—	—	—
Tomato	19	—	2	1	16	—
Potato	13	—	1	6	—	6
Subtotal	129	70	34	9	20	11
Other species	30					
Total	159					

The emphasis has been mainly on traits for herbicide, insect, virus & fungal resistance, and on product quality with a limited number of genes. Out of a total 159 releases. Argentina leads the way with 43, followed by chile, Mexico, Puerto Rico and the Republic of South Africa, with 17-20 each. India is far behind with just a few releases. There are fears relating to transgenic plants in society. These need to be addressed, not by brushing them aside, but by careful scientific analysis and educational programmes. All biosafety protocols must be followed rigidly. Let us deal with this issue more exhaustively now.

PUBLIC CONCERNS

The potential of biotechnology as a method to enhance agriculture productivity in the future has been accepted globally. However, because of its revolutionary nature, risk and uncertainty may be created by the process of genetic engineering and by the resulting genetically modified products. This may result from the introduction of a new unrelated DNA sequence into a recipient organism. The introduced DNA may have some unexpected effects on the cellular processes of the recipient organism. In addition to this, introduction of antibiotic resistance genes, as selection markers, pose serious implications in public health especially in genetically modified plants that are directly used for food purposes.

Risks are also associated with the genetically modified plants that are released into the environment. The nature of interactions with other organisms of the natural ecosystems cannot be anticipated without proper scientific testing. For example, modified plants with enhanced resistance to pests or disease threaten to transfer resistance to the wild relatives, which will have implications for biodiversity and ecosystem integrity. These, and many more, doubts plague the minds of common people. Therefore, it is very clear that if the biotechnology potential has to be completely realized, it has to be done in an extremely responsible manner. At each step, proper testing and scientific data have to provided. Not only this, proper regulatory and policy mechanisms have to be put into place.

It is therefore, absolutely essential, that conscious efforts be made especially by the developing countries to initiate well-defined programmes for the development and regulations of genetically modified plants. To make this possible and to benefically, ethically and sustainably reap the benefits of biotechnology, there is a pressing need for scientists, researchers, policy makers, NGOS, progressive farmers, industrialists, and representatives of the government, to come together on a common forum in order to discuss the common concerns and find solutions to them.

Will modified plants transfer their introduced genes into wild relatives growing nearby? Will modified plants that produce new compounds disrupt the 'balance of nature' in some way?

Could the planting of a restricted number of cultivators leave crop plants more susceptible to disease? Could the planting of a restricted number of cultivators lead to a reduction in biodiversity (of crop plants, weed species, insects and microflora in the fields in question)?

Will genetically modified plants be able to avoid the factors that regulate natural populations and thereby change the usual ' balance' between populations?

Similarly, there are concerns about the social and economic effects. For example, some of these concerns are:

- How will the structure of farming (particularly in developing countries) be affected by biotechnology?
- How will patent laws affect traditional breeders 'rights (*e.g.* the right to save seed from one year to the next)?
- Will plant breeding be left increasingly in hands of a few companies, and if so, what effects might this have?
- Will some countries be 'plundered' for their genetic resources?
- We also have a set of ethical and moral issues that need addressing. In particular,
- The consumer has the right. What will consumers be told about the new food products?
- Is it acceptable to 'interfere' with nature through genetic engineering?
- Do we have the right not to use all means available to improve crop plants, especially when so many people are under or malnourished?

Then there are several regulatory issues. Among these, the three key issues are:

- Do current regulations give sufficient protection to farmers, consumers, those who have invested in, and those engaged in research?
- Is there sufficient international legislation to ensure environmental protection?
- Do current regulations compromise the competitiveness of biotechnology companies by being excessively restrictive?

The Royal society of London had appointed a group of experts to examine various aspects including the scientific evidence concerning the risk of transfer of genes from Genetically Modified (GM) crop plants to wild species and non-GM crops, the uptake of genes from GM food by the digestive system, and the current state of the regulatory system.

The experts concluded that the chances of gene transfer happening are slight, provided the regulatory processes are followed, but that this must be kept under consideration.

The experts also concluded that the uptake of genes via the food chain is not a new issue because genes (*i.e.* DNA) are normal constituents of the human diet. Many products from GM Plants, such as sugar prepared from GM tomato paste, are so similar that they are regarded as 'substantially equivalent' others, for example flour from GM soya, may contain a new gene or its product, although many of the purification processes involved in food production will destroy and DNA present in the raw material.

Some GM foods have been produced using an antibiotic resistance 'market' gene, which is a laboratory device designed to identify genetically transformed plants.

The Royal Society has shown concern about the use of such genes in food products and it suggested that any further increase in the use of such markers in the human or animal food chain would be undersirable. Gm crop plants have been produced to improve insect tolerance and virus resistance, and to include herbicide tolerance, so that transferred to non-target species of plants, and that the development of resistance by target pests is minimised, the regulatory authorities must be assured that any negative effects would be no greater than those resulting from conventional procedures; and that any long term effects on the environment and ecology would be closely monitored, with statutory restrictions in place to control marketing; and also that best practice advice was adopted by growers.

The reliance on a case by case approach to the legislation, may result in lack of analysis of the overall impact of the technology on agriculture and the environment, and of the long term effects of gm. Although mechanisms are already in place to regulate many individual aspects of GM technology, there is no means for looking at GM technology as a whole. As independent, overarching, regulatory body is needed to span departmental responsibilities, monitor the enforcement of existing or future regulations, and strengthen the guidelines to growers of GM crops, such as those specifying the isolation distances between Gm and non-GM crops. The body should also review and monitor the membership of advisory committees and regulatory bodies.

BIOTECHNOLOGICAL APPLICATIONS RELEVANCE TO THE INDIAN FARMER

Increased food production must keep pace with the country's population increase. Not only the quantity of food made available should improve. To keep pace with the present population growth and consumption pattern, India's food requirement has been estimated to cross 225 million tons by 2005 AD. This would mean an annual agricultural growth rate of 6.7 percent, a daunting task considering the rapidly shrinking resource-base and fast declining input-use efficiency in major cropping systems.

Not withstanding the impressive gain in the agricultural production, the vast agricultural potential still remains highly under-utilized. There are serious gaps both in farm yield realization and technology transfer, as the national average yield of most crop varieties of most crop varieties is low.

Not many superior varieties is mot pulses, many oilseeds and vegetable crops have been evolved in recent past. In case of vegetables the varieties which were imported or developed many decades ago are still being grown by the farmers. Their yields are low and their inherent ability to respond to improve production technology is also limited. Farmers have accepted bybrids in my vegetable crops such as cabbage, tomato and capsicum.

RESEARCH AND TECHNOLOGY TRANSFER

The Central and State governments continued to invest heavilyu in research, hence the Indian public sector research system has increased in size in real terms but by mid 90s levelled off. The number of hybrids and varieties developed and released by ICAR Crop Improvement Projects and State Agricultural Universities in major and minor crops are enormous and have been well documented.

Private research to develop superior hybrids was initiated by few seed companies in mid 60s. By 1987-88, the number increased to over 12 and in 1997-98 it is estimated at 34. With the governments recognition of provate sector's performance capabilities and with "New Seed Policy" announced in 1988, the number of private companies involved in crop improvement programs have been steadily increasing (Table below).

NUMBER OF PRIVATE COMPANIES ENGAGED IN CROP IMPROVEMENT PROGRAM (1987-1998)

For want of financial, infrastructural and trained human resources, most private companies specialize in breeding superior hybrids in few crops characterized by low volume, high value and amenable to hybrid technology. In addition to the ICAR and Agriculture University research system, about 13 private sector seed companies are actively involved in vegetable R&D. Some of them have joint ventures with multinational or companies specializing in vegetable crop research and utilize research facilities available with their overseas partners.

Table : Number of Companies with Crop Improvement Programs

Crop	*1987-88*	*1993-94*	*1997-98*
Pearl Millet	12	23	25
Sorghum	10	12	15
maize	06	15	20
Sunflower	10	30	20
Cotton	09	21	34
Hibrid Rice	-	4	6

Few have created facilities for field testing imported and public bred varieties. The role of private sector in vegetable research is primarily in the development of F hybrids and few open pollinated varieties.

Table : Hybrid Vegetable and Melon seeds of Private R&D available in the market (1996-97)

Crop	*No. of Companies*	*No. of hybrid Varieties*
Tomatoes	11	45
Brinjal	10	38
Chilli	8	27
Bhendi	13	25
Cabbage	11	40
Cauliflower	8	21
Ridgegourd	6	8
Watermelon	5	13
Muskmelon	2	7

Number of private/public hybrids in major crops in market during 1995 are compared with the number in 1997.

In pearl millet, sorghum, maize and cotton number of hybrids introduced by private, companies has been steadily increasing. Many hybrids especially in paddy and cotton have regional adaptability. The number of hybrids which are accepted by the farmers and may be grown on 2 percent or more area is much lesser, so also the number evaluated in coordinated project trails or identified for release. Many hybrids recently supplied by the private sector embodies new technology which is either imported or developed by private companies. Private R&D's real investment in research has quadrupled between 1986 and 1997. Subsidiaries or joint ventures with multinational companies account for 30 percent of all private seed industry research in 1997. Both domestic companies and multinationals companies have started research programs on rice and mustard. Recently, Monsanto has also focused on hybrid wheat.

Table : Number of Private and Public Bred Hybrids in Major Crops in the Market during 1995 and 1997

	Number of Hybrids in Market				
	1995		*1997*		
Crop	*Private*	*Public*	*Private*	*Public*	*No of Private Hybrids grown on 2% plus area*
Pearl Millet	50	04	60	06	14
Sorghum	22	04	41	05	6
Maize	57	03	67	03	12
Sunflower	47	05	35	06	10
Cotton	73	15	150	15	19
Hybrid Rice	4	4	8	4	-

Few private sector companies have also initiated joint ventures (*e.g.* Mahyco-Monsanto Biotech, ProAgro-Agro-Evo) to transfer specific genes to their product lines to provide effective protection against pests. Some of them have taken initiative to treat their seed with high-tech chemicals to provide protection against sucking pests (aphids, jassids, thrips) during seedling and grand growth crop stages and diseases (downy mildew in bajra) during flowering.

COMMERCIAL SEED PRODUCTION

Majority of certified/commercial seed is produced under contract with trained seed multipliers in suitable agro-climatic conditions. Both public and private companies make contact with progressive farmers or trained seed multipliers and negotiate terms and conditions of production. Individual seed multipliers have small areas to contract and this greatly increases the cost of production and administration. To overcome this situation, contracts are arranged through local organizers, who in turn contract individual farmers, consolidate individual fields intoblocks and supply to the company the ultimate product.

Companies generallyprepare a technical package for seed multiplication specifying the kind of soil needed, isolation distances required, staggered plantion in case of a hybrid if needed, distances among rows and plants, depth of planting, time of nutrient and spray applications, frequency and time of irrigation, mode of harvest especially in case of hybrid seed etc. They prefer to contract between 300-400 acres in a village and designate such an area as a "seed village". They create seed processing facilities and place technically trained staff to guide seed multipliers, monitor actual production and supervise seed processing. Generally one such unit serves a cluster of 5-10 seed villages

Only seed lots meeting genetic Purity, Physical Purity, germination and disease standards prescribed by the government of India, are packed and commercially sold. User-farmers have become very conscious not only of genetic purity and germination, but also of physical attributes of seeds. They prefer bold, shining and plump seed, unaffected by mould The companies therefore, organize seed production in areas with appropriate agro-climatic conditions and no rains in the post-flowering period of seed crop. They also ensure proper cleaning to remove undersized, premature, broken seeds or inert matter on cleaner-cumgraders, gravity table, de-stoners and thus improve physical attributes of seed lots. Seeds are treated with recommended pesticides to prevent them against stored grain and soil pests. To add value to their seed, many companies now treat them with specific chemicals to provide protection agains sucking pests during seedling stages or specific diseases such as downys mildew for 40-45 days after emergence.

The seed production of most hybrid crops is generally undertaken in Western and Southern parts of the country because of congenial agro-climatic conditions and possibility of taking two seed crops in a year if necessary.

The prominent states are Gujarat, Maharashtra, Andhra Pradesh, Karnataka and Tamil Nadu. In contrast, the Northern states from Punjab to West Bengal are dominated by rice, wheat, and legume crops which are less profitable. Consequently there is a major difference in the cadre of seed industry between the North and West/South of India. Technology of seed production is moving at a rapid pace. Hence the companies have to ensure that they pass on the technology to the contract seed multipliers to harvest high yields of good quality seeds. Most public bred hybrids and variety seeds are certified while the proprietary hybrid seeds are sold under truthful label. All seeds meet minimum prescribed standards but for proprietary hybrids quality monitoring is through their in-house system.

PRIVATE HYBRID'S SHARE OF SEED MARKET

Private sector has played a major role in supplying quality seeds of both private and/or public hybrids. It is interesting to note that 69% of the pearl millet seeds produced during 1996-97 season was of private hybrids. Other crops where private research plays key role in seed supply are sudangrass sorghym (100%), sunflower (98%), Maize (90%), sorghum (52%) and cotton (50%). The contribution of private research in value terms is also steadily increasing. The share of research hybrids to total turnover in these crops was about 70 percent in 1996-97 as compared with 46 percent in 1990-91. Private hybrids are successfully meeting the regional needs of early maturity, tolerance to pests, unpredictable agro-climatic condition and soil variations.

GOVERNMENT INTERVENTION: LEGISLATION AND POLICY ISSUE

At the highest level, government policy is to reduce controls and to make the economic system more transparent in practice however, there is potential proliferation of legislation affecting seed sector. Given the current climate of economic development there is an urgent need for positive relationship between the government and the private sector. In order to fulfil farmer's needs and GOI food production targets set for the year 2005, the existing misapprehensions should disappear to ensure private sector investment in infrastructural improvement and expansion to increase quality seed production.

Presently Indian seed industry is governed by the following Acts:

- Seed Act 1966
- Seed Rules 1968
- Seed Control Order 1983
- Essential Commodities Act 1955
- Package Commodities order 1975
- Standrads of Weights and Measures Act 1976
- Consumer Protection Act 1986.

Export Regulations and Quarantine :

- Plants, Fruits and Seeds (Regulation of import into India) Order 1989
- Urban Land Ceiling Act 1976
- State Acts of land acquisition/land use
- State Acts for the control/movement of crops and seeds.

The policy framework hitherto has been regulatory with excessive legislation. Above plethora of legislation covering the seed sector is out of date and restrictive rather than progressive. The multiplicity of Acts, rules and administrative orders have resulted in over regulation of the nascent industry.

BIOTECHNOLOGY FOR SOCIETAL DEVELOPMENT IN INDIA

The application of biotechnology for societal development has always received special attention since it benefits socially and economically disadvantaged sections of the society. Through this programme a number of projects are implemented for economic and technical empowerment of the beneficiaries. Efforts are made through training, demonstration and appropriate R&D inputs for providing tailor made solutions for location specific problems of SC/ST, women and rural population using modern biology and biotechnology. The following pages provide the highlights of the achievements.

PROGRAMMES FOR SC AND ST POPULATION

Biotechnology-based Programme for SC/ST population has benefited around 15,000 people directly and about 45,000 indirectly through 54 projects during the year. The projects have been supported in the areas of cultivation of medicinal and aromatic plants, biofuels, production of biofertilizers and biopesticides, organic farming, aquaculture, mushroom cultivation as well as human healthcare interventions etc. The universities, public funded institutions, Krishi Vigyan Kendras and voluntary/non-governmental organizations have been involved in the implementation of the projects. The target groups have received the benefit of hands-on training and field demonstrations.

Vermicomposting: MR Morarka GDC Rural Research Foundation Jaipur implemented a project in Jalgaon district of Maharashtra, to benefit 2600 SC/ ST target population on economic utilization of waste, agriculture residues and cow dung. Over 1780 MT of waste, agricultural residues and cow dung was utilized for vermicomposting through establishment of 28 community facility centres.

Beneficiaries are utilizing vermicompost on crops like onion, sugarcane, cotton, banana, papaya and wheat. Some farmers switched over from conventional crops to cash crops such as colorful capsicum, gerbera as well as inter cropping of safed musli, turmeric and ginger in banana plantation. Farmers have been trained to use bio-fertilizer, foliar sprays and mulching to use the agro-waste for fertility and management. Marketing linkages have been established with "Organic Shoppe", through which farmers are getting 10-20% premium for their organic produce. About 20 entrepreneurs have taken up vermiculture as commercial activity. extension Material in the local language was distributed to the farmers.

Mushroom cultivation: A project on promotion & production of essential oil, mushroom and vermicompost was implemented by RRL, Jorhat for revenue generation of the ST population of Arunachal Pradesh. Motivation and demonstrationcum-training programmes have been conducted in 13 villages of five districts on cultivation of citronella, production of mushroom and vermicompost. Demonstration units have been established at RRL Branch Naharlagun farm. Mushroom spawn, earthworms and citronella saplings were distributed to the beneficiaries. Several beneficiaries established vermicompost units on commercial basis and are selling vermicompost. The beneficiaries at Ziro, Jairampur, Doimukh, Manmao, Hetlong and Tengman have established oyster mushroom cultivation units. So far, 10 acres of land have been planted with citronella and 7 acres of land with patchouli at Sonajuli village of Papum Pare district.

Medicinal plants: A project on strengthening health care practices and conservation of medicinal and nutritious plants was continued at Herbal Folklore Research Centre, Tirupati for implementation in 100 villages of two Mandals in Andhra Pradesh. Training workshops were organized with the help of Yanadi tribal people for community health volunteers on identification and sustainable harvesting methods of 20 species of locally used medicinal and aromatic species. The reputed healers were encouraged to prepare medicines needed for PHC from the identified plants. About 6000 seedlings were distributed to the beneficiaries for planting in kitchen gardens. Green health kits (samples of herbal medicines) and educational material in local language containing identification, cultivation techniques and preparation of medicines with their therapeutic properties were distributed to the trainees.

A project on creating awareness and promotion of medicinal plants for the production and marketing was implemented by RRL, Jorhat to benefit ST population of Arunachal Pradesh. Awareness and training programmes were organized on the importance, scientific cultivation and marketing of Piper longum, Kaempferia galanga, Andrographis paniculata, Stevia rebaudiana, Rauvolfia serpentina and Mentha piperita. Three nursery-cumdemonstration centers were established at Roing (1000 ft msl), New Bomdila (8160 ft msl) and Ziro (4350 ft msl) and 19000 saplings were distributed to ST farmers/NGOs for introduction and propagation in their fields. At high altitude, villagers were advised to cultivate Aconitum heterophyllym, Coptis teeta and Picrorrhiza kurroa. Beneficiaries have stopped collecting wild Homalonema aromatica, an important but rare aromatic plant of Arunachal Pradesh, and have started its cultivation in their farms. Ten community nurseries have been established with the help of Vivekananda Trust, Roing at ten villages of Lower Dibang Valley District. Twenty farmers were trained on cultivation and post-harvest management of medicinal plants and marketing tie-up has been established. Around 1266 local ST people participated in the awareness programmes.

A project on organic cultivation of medicinal and aromatic plants for income generation and sustainable development in Uttarkashi District of Uttaranchal was undertaken through Forests Conservation Programme of World Wide Fund for Nature, New Delhi. Demonstration-cum-Resource Material Production Centre established for training on vermicomposting and cultivation of species.

Construction of poly house, vermi-composting units and planting of seeds of Rauvolfia serpentine completed. Awareness and motivation camps were organized for the farmers. Extension material brought out in the local language for cultivation of Rauvolfia serpentina, Coleus barbatus and Pelargonium graveolens were distributed to the farmers.

A project on developing skills and economically sound vermicompost units utilizing distillation waste of Mentha arvensis (menthol mint) was implemented for the benefit of SC/ST communities of Lucknow and Barabanki Districts of Uttar Pradesh. Around 200 farmers were trained on menthol mint nursery raising, transplanting, distillation and practices followed in vermicomposting. Demonstrations were carried out in farmers' field using high yielding variety 'Kushal'. Farmers trained on organic farming started vermicompost production using distillation and other agro-waste. The yield data obtained from the demonstrations showed the possibility of reduction of chemical fertilizers by 75% without affecting the yield.

BIOFUELS

A project was continued on cultivation of Jatropha in arid lands of Namakkal District in Tamil Nadu for production of biofuel and other products for SC/ST community by Centre for Research in Social Sciences, Coimbatore. A high tech nursery was established and plantlets of Jatropha were raised. Training programmes were organized for 223 members on cultivation of Jatropha in 31 acres of land. About 30,000 plants were supplied for cultivation. Awareness camps, training and demonstrations were conducted on Jatropha cultivation, oil extraction and soap making. SHG groups were trained in nursery techniques. Women self help groups were trained in various activities of jatropha cultivation and soap making. Extension material bought out in the local language was distributed to the beneficiaries.

Biocontrol agents: A project was implemented on development of eco-preneurship among the coastal community of Pondicherry by MS Swaminathan Research Foundation, Chennai. Sixteen SHGs involving women and youths were formed in Ariyankuppam, Nonankuppam, Kakayanthope and Veerampattinam villages. Demonstrations and trainings were conducted to educate trainees on basic aspects of IPM, importance of biocontrol agents, production and marketing of Trichogramma, importance of human nutrition, role of edible mushrooms and production & marketing of oyster mushroom. Micro-enterprise was developed for mushroom and Trichogramma production. Technologies were disseminated on biopesticide (Trichogramma), oyster mushroom production and vermicomposting to recycle the spent waste of oyster mushroom. Extension material brought out in the local language was distributed to the villagers.

AQUACULTURE

A project was undertaken on enhancing livelihood of SC/ST people in Orissa and Tamil Nadu by MS Swaminathan Research Foundation, Chennai for introducing various aquaculture enterprises to 250 participants in backyard ornamental fish culture and carp seed production. Fingerlings produced were distributed to the entrepreneurs for stocking. Various aspects on ornamental fish breeding, carp nursery and gold fish culture were disseminated to the self help groups, which helped them in earning a revenue of Rs. 200/month/

household and an income of about Rs. 3200/from carp nursery and Rs. 8500/ from vegetables and flowers as bund crops. Community ponds were effectively utilized by stocking with rohu, catla, mirgali, Chinese & Indian carps and marketable size fish was harvested. Technologies transferred to the entrepreneurs on community pond aquaculture, carp nursery and ornamental fish culture and extension material in local language were distributed to the beneficiaries.

A project on utilization of integrated marine living resources by SC youth from Tuticorin region of Tamil Nadu was undertaken by College of Fisheries, Tuticorin. Lobster fattening and crab culture techniques were standardized and demonstrated to the beneficiaries, who have grown lobsters in cages for about 4 months and produced 250-350gms size for marketing. The beneficiaries were helped in availing credit facilities and bank loan. In the project around 120 people of target population have been trained. A manual on crab and lobster fattening technology was brought out for the distribution to the beneficiaries.

A project on freshwater prawn culture was implemented in village Naubasta Kalan in Allahabad District by University of Allahabad. Demonstrations were conducted on prawn hatchery technology and standardized methods for biocontrol of prawn disease for integrated giant freshwater prawn (Macrobrachium rosenbergii) culture. Trainings were conducted on scampi culture, breeding and seed production and microbial inputs were utilized for feed silage preparation using indigenous technology. Apiculture activity was introduced and disseminated to the target population and extension material was distributed to the beneficiaries.

GENETIC COUNSELING

Tribal population of northeast region was screened for anaemia and nutritional deficiencies through a programme implemented at Vivekananda Institute of Medical Sciences, Kolkata. The tribes of Arunachal Pradesh, Assam and Tripura displayed high incidence of anaemia. Awareness programme was organized for school children on inherited anaemia in pre-marital and pre-natal population. More than 1500 families from 80 villages were educated on anaemia, thalassaemia and hemoglobinopathies and counseling was provided to them. Leaflets were distributed for awareness of anaemia with special reference to Thalassaemia in local languages.

A project on molecular epideomiology of alcoholism was implemented jointly by Centre for Cellular and Molecular Biology, Hyderabad and Tribal Research Centre, Udagamandalam, Nilgiri District to study Kota tribe population in Nilgiri hills. General health awareness camps were arranged in seven villages for periodical medical check up. Blood samples were collected from 620 individuals and clinical data on health profile revealed that the disease morbidity is more for hypertension, heart related problems and diabetes. Genealogies indicated that the alcohol drinking habit is running in the families for the past two to three generations. Study undertaken on the health awareness camps and medical camps were found beneficial in counselling the tribal population.

FLORICULTURE

Over 180 women from Shillong, Itanagar and Chandigarh were trained in cultivation of commercially important and hybrid species of orchids. A total of 118 women of Shillong have raised 6 species *viz.* Dendrobium fimbriatum var. oculatum, D. longicornu, D.lituiflorum, C. pendulum, C. aloifolium and C. giganteum in their backyard. These plants have started blooming. The beneficiaries have started selling their harvest in local market. Through S&T Council, Bhopal 200 women were trained in cultivation of gladiolus, marigold, mogra, rose and tuberose.

FOOD PROCESSING

At College of Horticulture, Vellanikara, Thrissur, 88 women were trained in various aspects of entrepreneurship in preparation of cocoa products including molding, packaging and marketing. Twenty-two women have already set up their production unit and are selling their products in local market.

In another project at Fisheries College, Tuticorin, 500 women were trained in cook-chill technology for seafood and various parameters for quality assurance *viz.* estimation of moisture content, bacteria load, presence of bacterial species, the nutritive value, shelf-life etc. The trained women were grouped in 48 SHGs for establishing market for seafood products.

At the Central Food Technological Research Institute, Mysore data on biochemical and microbial analyses for nutritionally important factors like protein, amino acids, isoflavones, genistein, diadzein and methods for improvement of shelf life were collected on for 'Tofu' preparation. Also a byproduct of Tofu preparation, Okara, was found to be rich in protein and amino acids has been used as a food supplement. In another project at CFTRI, 17 women were trained in preparation of various food products of high nutritive value, methods of removal of anti-nutritional factors like phytic acid and trypsin inhibitor found in soya bean, better texture and improved shelf life. The food items included fermented products of soya bean and fruit pulp. Also 34 women were trained in spirulina production and its application as health food. Through CFTRI, technologies for rural bakery are being disseminated to rural women in three villages.

POULTRY AND LIVESTOCK FARMING

The 'Action Research on Health and Socio Economic Development', Bolangir, Orissa, identified 100 women and trained them in poultry rearing.

The beneficiaries were provided with parent birds of Carigold (dual purpose) and Vanaraja varieties. The beneficiaries were able to earn Rs. 1000/per month through the sale of eggs and birds. The beneficiaries were also trained in feed formulation using local ingredients, management practices and entrepreneurship skill. The trained women formed SHGs and are approaching a bank for setting up their poultry farms. The market is established through state owned apex cooperative known as OPOLFED.

Through three projects supported at Tamil Nadu Veterinary & Agricultural Sciences University and its Extension Centres at Namakkal, Chennai and Tiruchirapally, over 1200 women were trained in poultry farming including feed formulation, farm management / health care and preparation of various food items such as quail egg pickle and ready-to-eat food products etc.

Aquaculture: Under a project at the North Eastern Hill University, Shillong, techniques were standardized for induced breeding for common carp and 4,00,000 fish seeds were reared up-to fries for distribution to the beneficiaries. Selected 92 women were trained in fish farming techniques including water quality testing, pond maintenance, induced breeding, harvesting and feed formulation using slaughter house waste. They were provided with fish fries for rearing in their own pond.

Over 470 women from Bhopal, Shimla, Solan and Bhubaneshwar were trained in polyculture fish farming. Interested women were also provided advance training in composite carp culture, integrated fish farming with duck/ poultry, pond management and value added products including fish processing. These women are selling their products in local market and are earning Rs. 3000/per harvest/family.

Considering the opportunities in ornamental fish, a project was supported at the University of North Bengal. The identified women, 57 from Darjeeling and 54 from Jalpaigudi district were trained in rearing and breeding of guppy (Poecilia reticulate) in cemented cisterns and angel (Pterophyllum scalaer) in ponds. The women are being trained in management skill and marketing.

ORGANIC FARMING / VERMICOMPOST

At the Sri Sathya Sai Institute of Higher Learning, Anantapur Campus, Andhra Pradesh, neem varieties with higher level of azadirachtin were identified in 5 villages. Thirty trees were selected and 40 women from two villages were trained in fruit collection, drying and storage. Specialised training on maintenance of neem processing gadgets, processing of neem fruits for production of cake, oil or powder, maintenance of aza-rich plantlets, hardening, harvesting and application of neem based biopesticide was provided to five resource persons

At the University of Agricultural Sciences, GKVK Campus, Bangalore, five villages of Bangalore and Kolar were surveyed for the availability of organic waste. The waste materials were mostly sugarcane trash, coconut coir pith and vegetable wastes. Demonstration programmes for 20 to 25 farmwomen from each identified village were organized. Six interested women were given onfarm training. The technique of Trichoderma enriched vermicomposting was perfected and demonstrated at University campus.

At MSSRF, Chennai, phosphate-solubilising local strains were collected and screened. Simultaneously several awareness programmes on application of biofertilizer were organized for the villagers of Rediarchatram block. About 300 interested farmers were trained in large-scale production of biofertilizer.

Under the project support at Dangoria Charitable Trust Hyderabad, apart from training women in processing various items of food, 100 women from 13 villages were trained in production of vermicompost, neem-based biopesticide and their application in organic farming, About 100 vermicompost units were set up using the ring method. The beneficiaries produced over 130 quintals of compost. The beneficiaries were encouraged to raise nurseries of drumstick, papaya and creeper spinach at their backyard using vermicompost. The beneficiaries were able to sell excess vermicompost and planting material.

HORTICULTURE

Through the Tamil Nadu Agricultural University, Coimbatore, 54 progressive farm women were identified for training on cultivation of hybrid rice. The women were provided with quality seed their cultivation over an area of approximately four acre. The yield for the first harvest (December to May) was recorded at 2000 kg/ha. The beneficiaries were able to earn an additional income of Rs.20,000/per acre/year through three crops in an year. At Supi farms, Uttaranchal non-traditional vegetable crops were grown on farmers' land in a small way. An herbal garden was established. Incidence of pests to various crops was evaluated and farmers were trained on production of vermicompost and application of neem based biopesticide in horticultural crops.

MUSHROOM CULTIVATION

In an attempt to empower rural women through transfer of mushroom cultivation technology 93, women were trained at Himachal Pradesh Krishi Pradesh Vishwavidyalaya, Palampur and 750 at Tropical Botanic Garden & Research Institute, Kerala in mushroom cultivation. In addition, 15 women were trained in spawn production at TBGRI. The beneficiaries were given 15 days training in cultivation of Oyster and Agaricus species using wheat straw. The first harvest of about 1000 kg was sold @ Rs. 16/kg. The beneficiaries were also trained in various recipies of mushroom. Market linkage has been established for their products.

Under the project support at Dangoria Charitable Trust Hyderabad, apart from training women in processing various items of food, 100 women from

13 villages were trained in production of vermicompost, neem-based biopesticide and their application in organic farming, About 100 vermicompost units were set up using the ring method. The beneficiaries produced over 130 quintals of compost. The beneficiaries were encouraged to raise nurseries of drumstick, papaya and creeper spinach at their backyard using vermicompost. The beneficiaries were able to sell excess vermicompost and planting material.

SERICULTURE

A total of 550 women were trained for various activities including the organic cultivation of mulberry & rearing of cocoon and value added products from waste at the Central Sericultural Research & Training Institute, Mysore and at Vidyasagar University, Midnapore, West Bengal. The beneficiaries are getting an additional income through the sale of value added products and surplus vermicompost.

HEALTHCARE

Kidwai Memorial Institute of Oncology, Bangalore, Karnataka organized over 50 awareness camps on Carcinoma cervix to educate 2071 women for its prevention. Blood samples collected from 1830 women based on their medical history, 1812 were found negative and 18 positive cases were referred for treatment. A project on mental retardation including awareness generation among rual women, screening for early detection and genetic counseling was supported to Manovikas Bio-Medical Research. Over 7,000 families from three Gram Panchayats were surveyed. Ninety-one of these showing signs of mental retardation were shortlisted. The families were informed about the possible risk factors in future. Analysis of blood samples from these for genetic carrier showed only 4 cases of low level of fragile X chromosomes.

A project is supported at Saraswati Dental College, Lucknow for dental fluorosis. Water samples from 12 villages around Lucknow and Kanpur were analysed for fluoride content. Health camps were organized in 4 villages with a team of dental surgeon. A total of 2860 people were examined for dental fluorosis. The severely affected persons were advised for regular dental examination at SGPGI, Lucknow. They were also provided with calcium and cholicalciferol supplement. The blood samples from the severely affected boys and girls within the age group of 15 years and pregnant women were examined for their blood calcium level. Under another project supported at Mahavir Hospital And Research Centre, Hyderabad, Andhra Pradesh, blood lead levels in children and adults including men and women were measured to study environmental genotoxicity in women and children occupationally exposed to lead, a heavy metal.

Index